PROFITABLE FARMING NOW!

Edited by Mike Brusko
and the editors of *The New Farm*
George DeVault
Fred Zahradnik
Craig Cramer
Lesa Ayers

Regenerative Agriculture Association

Design and Art Direction: John Lotte
Associate Art Director: Ardyth Cope
Cover Photo: Mitch Mandel
Cover Design and Art Direction: John Lotte
Production Coordinator: Barbara A. Herman
Composite Typesetter: Brenda J. Kline
Typesetting Composition: The Type House

Copyright 1985 Regenerative Agriculture Association.
All rights reserved. No part of this publication may be reproduced or transmitted in any form or by any means, electronic or mechanical, including photocopy, recording, or any information storage and retrieval system, without written permission from the publisher.

Published in the United States of America by the Regenerative Agriculture Association, 222 Main Street, Emmaus, Pa. 18049.

Library of Congress Cataloging in Publication Data
Main entry under title:

Profitable farming now!

   Bibliography: p. 104
   1. Organic farming. 2. Farm management. I. Brusko, Mike.
II. Regenerative Agriculture Association (Emmaus, Pa.)
III. New farm.
S605.5.P76   1985   631.5'84   84-26271
ISBN 0-913107-01-8

Made possible in part by a grant from the Rockefeller Brothers Foundation, and by RAA membership contributions.

CONTENTS

Your Best Friend **5**
When you think about it, the only person who *really* cares about your farm—and has the power to ensure its success—is you.

'Where Do I Start?' **7**
Even the journey to more profitable farming has to begin with a first step. Here are some thoughts from farmers who have already taken it.

What The Research Reports Haven't Told You **20**
Straight answers to some of the most often asked questions about reducing farm chemical costs.

The Dollars And Sense Of Resource-Efficient Farming **29**
The Rodale Research Center's Conversion Project. How we did it, what we learned, and what it means to you.

When In Doubt, Ask A Farmer **46**
We did. And surprisingly few said their yields and income fell when they cut back on chemicals.

They're Already Farming More Profitably **55**
Meet nine farmers who took charge of their farms years ago, and learn how their ideas can work for you.

    They Saved $60,000 On P And K 55
    Good Yields For 10% Less 57
    He Cut His Chemical Bill In Half 62
    Forget Foxtail 64
    He *Nets* $60,000 A Year—
      Without Buying Fertilizer 67
    This Corn Yield Champ Doesn't
      Use Herbicides 69
    Why Grow Corn? 70
    Build A Nitrogen-Planting
      Weed Killer 73
    He Plows Down 100,000 Pounds
      Of *Homegrown* N 74
    Reversing Erosion 77
    18 Years Of Top Yields—Without P And K 79
    Who Needs Herbicides? 80
    Grow Less And Make More 84

Appendix **85**
    Regenerative Agriculture Resource List 85
    For Further Reading 90
    People Who Can Help 95

Index **99**

Your Best Friend

An Introduction

You hear a lot of talk these days about commodity prices, how low they've sunk and who is to blame. Republicans say it's all Jimmy Carter's fault. His Russian grain embargo wrecked the world grain trade, ruined the reputation of the United States as a reliable supplier, and vaulted countries like Brazil into the lead of grain exporters. Democrats say it's not all that simple, but they don't have much more of a comeback.

You hear even more talk about what must be done to restore agriculture to economic health. Nearly all politicians call for increasing productivity, adding value to our ag exports, jealously protecting our remaining export markets and conquering new foreign markets. If the latter requires dumping or other unfair trade practices, some say, then so be it. Farm groups, as they've been doing for years, variously demand full parity, guaranteed prices and other government intervention. At the same time, the White House is considering taking a meat cleaver to what it considers runaway farm price and income supports. The fertilizer industry sees the answer in higher yields caused by higher fertilizer sales. Never mind that bumper crops drive commodity prices even lower. Herbicide makers, quick to recognize soil conservation as a mighty marketing tool, push "conservation tillage" as the cure-all for soil erosion—and sagging chemical sales. Never mind the growing health and environmental costs of that. Just consider herbicide bills that now run as high as $50 per acre. Enough said. Such a list could go on and on.

Daylight Ahead

Despite all that, daylight is starting to show at the end of the long, dark tunnel now suffocating agriculture. The light grows a little stronger every day. I'm extremely optimistic about the future of farming. There has never been any doubt in my mind that people—profit—and biological permanence can be returned to farming. It's happening already, one farm at a time, throughout North America.

Yes, we can and will regenerate our agriculture, but not by following the route mapped out by some of the special interests trying to pass themselves off as the best friend agriculture could ever hope for. Case in point is genetic engineering or biotechnology, which a growing number of people see as the cure for whatever ails agriculture, now and in the future.

"In some cases, seed sales will be linked (through advertising) to chemical packages," according to University of Kentucky sociologists Lawrence Busch, Michael Hansen, Jeffrey Burkhardt and William B. Lacy in their 1984 paper "The Social and Scientific Impacts of the New Plant Biotechnologies." The paper says that the real needs of farmers are already taking a back seat to the vested interests of "seed companies and the chemical companies that run them. Neither scientists nor administrators appear aware of the potential conflict between the interests of farmers and those of agribusiness. In a very real sense, the new biotechnologies will make farmers more dependent on industry in a way that they have never been before," the sociologists conclude.

"The potential for farmer-industry conflict is manifested in recent Senate testimony given on behalf of the American Farm Bureau Federation. The Farm Bureau argued that public research should be focused on reducing the cost of inputs, particularly chemicals. Conceivably, the new biotechnologies could be used to do just that. All the evidence, however, suggests that just the reverse is happening. And, it is clearly advantageous to the chemical companies to sell more rather than less chemicals."

Farming's Best Friend

Who can help reverse such trends? Who can put agriculture on a steadier course, one that relies on information, management and renewable natural resources, rather than purchased products? In short, who is farming's best friend? Your best friend?

You are.

You're the new farmer, the person who is unafraid to look for a better way. And you're finding it, proving that new ways of farming are healthier, safer and more profitable.

And that's what this book is all about. This is your story, the stirring saga of how farmers just like you are doing nothing short of reinventing agriculture. It's history in the making, an event that agribusiness just can't ignore. Once, independent-minded farmers like you were thought of as prodigal children who would one day return to your senses and go back to farming by the book. Now, the experts in industry and government are taking you seriously. They have to because you're not "farming the way grandpa did," as the critics like to say. You're taking the very best technology has to offer, tempering it with the hard realism that comes from making a living on the land, and leading farming toward a brighter future.

In this book, you'll meet dozens of your fellow pioneers. These pages explain how they're devising new, cost-effective ways of controlling weeds, rebuilding the soil and safeguarding the environment, while producing good yields. Their triumphs are presented as well as their failures, because mistakes are often the best teacher around.

"Profitable Farming *Now!*" was rushed into print in near-record time because agriculture needs solid answers right now. No longer can we afford to wait years for laboratory-approved practices that may or may not benefit the farmer. What farmers really need now is practical cost-cutting information. And there is no better source of that than farmers themselves. Of course, no one has *all* the answers. No one ever will. But there are a lot of proven answers and good ideas in this book. Most of them come from farmers who are using their own resourcefulness to put people—*profit*—and biological permanence into their farming, now. Maybe you're already using some of these methods, too. Or, maybe you're about to. The number of farmers rethinking the way they do business is growing daily. People see you working to solve the same problems they face. They know you're on the verge of something new and great, and they're following your bold lead.

Yes, the new farmer is a force to be reckoned with. You're taking charge of your farm and your destiny, showing that you are the best person to be entrusted with the major management decisions that were taken away from most farmers years ago. No longer will an uncaring system dominate the way you live and farm.

True, there is still a long way to go. Many farmers face staggering debt loads, for example. Some are simply too far in the red to survive. But more and more farmers like you are already making tremendous progress. Just imagine how much more you can accomplish by working together—with your best friend on your side. Keep up the good work, friend.

George DeVault, Editor, *The New Farm*
January 9, 1985
Emmaus, Pennsylvania

'Where Do I Start?'

Even the journey to more profitable farming has to begin with a first step. Here are some thoughts from farmers who have recently taken it.

"I'll be honest with you, it's a very scary area for me. You've got to be a viable business."

To any farmer who's even thought of giving up chemicals, Jim Greuel's words are like a new pair of work boots: uncomfortable at first, but maybe—just maybe—worth getting used to.

Greuel farms 480 acres just south of Mattoon, Ill. And like many of his neighbors, he's becoming increasingly concerned about the long-term sustainability of his operation. "I've been growing corn and soybeans for the last 20 years," he says. "But there are going to have to be some changes. I can see the erosion and compaction. And 20 years is just a pinhead in time."

Yet in summer 1984, when Greuel finally decided to do something about the problem, he found himself with far more questions than answers. Would he have to take land out of production? Would the landlord from whom he rents about 100 acres accept the idea of fewer chemicals? What would happen to his yields? Could he really control weeds without herbicides? What kind of rotation should he use?

This type of self-examination is not only common, but essential during the early stages of such a change, says Dr. William C. Liebhardt of the Rodale Research Center (RRC). Having worked closely on RRC's five-year, chemical-to-organic "Conversion Project," Liebhardt has identified a number of cost-cutting management options farmers can put to work immediately. But he stresses that, before using any of them, a farmer must study two things very closely: his operation and his personal goals. "You try to find out as much about the operation as possible: the size; the location; what he's growing; whether he has livestock; his soil type, especially the drainage; and what the farmer thinks is possible in his or her situation," Liebhardt notes. "They're going to have to grow into this new situation.

"It's really at a personal level for the farmers," he adds. "If they're concerned about their profitability . . . their health . . . their environment, they've already started the process mentally."

For Greuel, reducing herbicides didn't appear to be a practical move right away. His flat, silt-loam soil drains poorly, and the traditionally wet springs in south-central Illinois make early cultivation risky, at best.

But the fact that his soil tests have long shown sufficient levels of P and K got him thinking. "I can see we've got the availability there, if we can just manage it," Greuel says. "All we've got to do is break it loose."

To handle the job, Greuel chose buckwheat. The plant's extensive, yet shallow root system is known for loosening compacted soils and making them more friable. What's more, University of Minnesota scientists say it's capable of using phosphates that are unavailable to other grains. That, and buckwheat's rank, weed-shading growth led Greuel to plant 120 acres of it in early August '84.

He plans to combine the buckwheat in fall. If he can get into the fields by May 25, he'll follow it with corn. Otherwise, he'll plant soybeans. "In this time of high energy (costs), I've decided I don't want to fight it and plant corn if it's too late," he explains.

At this point, Greuel is mainly interested in improving the organic matter level of his soil—an important goal, says RRC's Liebhardt, but neither critical nor realistic in the short run. "It's the constant addition that's important," he notes. "To build up the organic matter *substantially* is really tough."

Greuel recognizes that, so he's also designing a three- to four-year crop rotation that he hopes will reduce his need for chemicals within just a few years. "I'm getting back to a rotation with wheat and legumes, which will be about one-third of the rotation if we can manage it," he says. "In no way do I want to come along with soybeans back to back, or corn. You have weed problems, insect problems. The thing I see is to keep more green growing on the ground, even if it means less money."

The rotation will start with corn. Soybeans, either no-tilled or planted conventionally into chisel-plowed and disked ground, will follow. In fall of the second year, Greuel will plant wheat immediately after soybean harvest. He'll seed down the wheat to clover "anytime between February and May."

Greuel isn't sure what he'll do with the clover after wheat harvest. He says he may be able to take a seed crop later the same year. But more probably, he'll cut the clover for hay, then combine seed before resuming the rotation with corn.

Once the rotation is established, Greuel feels he'll be able to reduce his herbicide and fertilizer rates substantially. And he's quick to point out that, "for two years out of the four, I won't be using them at all, anyway. A little bit less is better, even though we will be using them otherwise."

Meantime, Greuel is minimizing his use of superphosphate, and relying more on rock phosphate for P. He's also experimenting with 40 acres of hairy vetch as a combination seed crop and nitrogen-supplier for corn. And though he still buys some nitrogen fertilizer, he no longer uses anhydrous. "I sold my ammonia applicator this spring," he says.

Greuel's problems are typical of many cash-grain farmers, and his proposed solutions appear well-supported by research. "Our rotation (RRC's Conversion Project) clearly supports what he's doing," Liebhardt says, "because you plant corn around the middle of May, and soybeans two or three weeks later." By then, he points out, warmer weather will give the cash crop a better start, and less rain makes it easier to cultivate.

If You Can't Replace, Reduce

"The farther north you go, the less flexibility you have with respect to your date of planting and mechanical weed control," Liebhardt adds. "It's just like our farm. We've got some fields we can cultivate two or three days after a rain, and others where you have to wait five or six days. Now, on the weed growth, something he should consider is banding (herbicides) over the row," he says. "He should take a worst-case scenario with respect to rain, and select some fields where he knows the chances of cultivating are still pretty good." Then, cut back on herbicides, or eliminate pre-emergence sprays, and cultivate instead.

Mulch/planter leaves crop seeds in a 4- to 5-inch-deep trench. Later cultivations clean row middles and feed soil back to the row to smother weeds.

Middlebusters near top of subsoiler shank on Cole Mfg. Co.'s mulch/planter throw the top 4 to 6 inches of soil, along with trash, stubble, insects and weed seeds, into the row middles.

Cal Huge uses a six-row John Deere field cultivator instead of pre-emergence herbicides in soybeans.

Cal Huge, who farms 600 acres near Eastover, S.C., took just that approach when he began reducing herbicide use back in 1979. Huge now grows soybeans without pre-emergence herbicides, and sometimes even without post-emergence herbicides. And he has trimmed his production costs by up to $23.59 per acre in irrigated, full-season beans, and by $25.50 in double-cropped (following wheat) beans.

The process begins at planting, with a six-row, Cole Manufacturing Co. mulch/planter. This unit is similar to a conventional subsoiler/planter, except that it has a "middlebuster": a wing-shaped, steel device near the top of the subsoiler shank that throws the top 4 to 6 inches of soil, along with trash, stubble, insects and weed seeds, into the row middles. This leaves the crop seeds in a 4- to 5-inch-deep trench after planting. Later cultivations clean the middles and feed soil back to the row to smother weeds. (For more details on the Cole mulch/planter, see "New Machines Turn Trash To Cash," *The New Farm,* May/June 1982.)

Ideally, if weather and time allow, he'll use a six-row sweep cultivator on full-season beans beginning the first week of May, then come back toward the end of May or early June, then at 10-day intervals until lay-by. The crop usually requires three or four cultivations. If Huge can't cultivate when he needs to, he turns to a post-emergence herbicide such as Blazer. He also mulch-plants and cultivates his corn.

In the long run, though, farmers and scientists agree that one or more years of a legume sod in the rotation is an even better way to start a farm on the road to higher profits. In addition to its use as a forage, hay or seed crop, it will help disrupt weed cycles and supply inexpensive nitrogen to cash crops that follow. Provided, of course, that the farmer selects an appropriate legume species for his area, and pays attention to soil pH. "You'll know by looking at the field whether you're going to get the nitrogen and the weed control benefit you're looking for," says Liebhardt. "If you've got a poor hay stand, that (soil pH) could be the problem. Or, if there's a significant amount of winter-killing in your hay, you may have to get a different species."

Soil pH should be checked and adjusted by liming, if necessary, to bring it up to between 6.5 and 7.0, where most legumes flourish, he adds.

Moellring uses a Noble ridge-till cultivator to control weeds in corn and soybeans. "You can set the disk hillers to push that dirt right around the rows," he says. "And you know that dirt's going to kill a lot of weeds."

Small Grains, Legumes Restore Depleted Soil

Like Greuel, Mel Moellring's biggest concern was crop yields when he decided to use non-chemical methods on the 155 acres he added to his western Illinois home farm in 1980. His rolling, clay-loam soils had been in continuous soybeans for 10 years, and were row-cropped and fall-plowed for at least 20. "How long will this conversion take?" Moellring asked himself. "Will it fall flat on its face? Should I work into it gradually?"

Moellring opted for the latter. In early fall 1980, he planted about 30 of the 155 acres to wheat and orchardgrass, and seeded that to red clover the following spring. In '81, he combined the wheat and sowed the rest of the acreage to corn and soybeans. Moellring fertilized with anhydrous ammonia, and used recommended herbicides for weed control. But he applied no insecticides. "I'm really not in the habit of it," he says. "One field was pretty well eaten up with rootworm, so I just didn't have a good stand. I figured the next year I'd take care of them with the rotation."

The following year, Moellring sowed an additional 30 acres of wheat to clover, this time applying about 7 tons of cattle manure per acre. Fifty-five acres of corn and 40 acres of soybeans also received manure, but Moellring didn't reduce his fertilizer rates on these crops, yet. "I wanted to get back the organic matter more than the nutrients," he explains.

Herbicides were still needed, too.

After harvest on some of his 1982 corn and soybean fields, Moellring planted a mixture of about 1.5 bushels of wheat and 2 to 3 pounds of orchardgrass or Timothy per acre, seeded to 8 pounds of red clover in mid-March.

In '83, he mowed 30 acres of wheat for extra forage, and combined the rest. But it's the clover that really began making the rotation profitable. Moellring earns about $240 per acre from a clover hay cutting in June, then harvests seed from the same fields in late August or early September. With yields ranging from three to five bushels per acre, and prices averaging $60 per bushel, he says the seed crop earns roughly $300 per acre.

Saving seed for future plantings means Moellring's only cost for the crop is combining. "And that's in the family," he says with a grin.

Third-year clover is grazed in spring, then double-disked for the following year's corn. "But I think you get more nitrogen if you let the clover grow that third year, instead of grazing it," Moellring says. "And I won't have any weed problem here, either, because it's the first year out of the legume."

By 1984, Moellring was able to reduce or eliminate chemical use in nearly all of the fields at the 155-acre site. Twenty acres of soybeans, which had followed corn, received a small amount of herbicide, but no manure or fertilizer. Another 30 acres of beans, planted into a torn-up hayfield, received neither chemicals nor manure. Likewise, a 20-acre cornfield that had followed three years of clover "got extra manure," but no chemicals.

Moellring credits a four-row Noble ridge-till cultivator, which he's borrowing from a nearby equipment dealer, with letting him control weeds with fewer herbicides. He says the machine throws much more soil than the field cultivator he'd been using. And its large, heavy-duty sweeps, made for throwing trash, weeds and crop residue out of row spaces, are capable of shearing tougher weeds. "I think I'm going to buy it," says Moellring, quickly adding that ridge-planting is definitely not in his immediate plans.

Overseeding clover into wheat helped Mel Moellring eliminate chemicals on most of his land. Clover hay and seed earn about $540 an acre, he says.

"You can set the disk hillers to push that dirt right around the rows. And you know that dirt's going to kill a lot of weeds. If I bought it and used no herbicides for two years, it would be paid for on my acreage."

Meantime, Moellring uses livestock manure and a corn-corn-soybeans-small grain/legume hay (two years) rotation to keep chemical costs low on his 320-acre home place. In fact, he says, about the only chemical he can't do without is grass herbicides in soybeans. Rain and a lack of time generally prevent him from getting good mechanical weed control. "I just haven't mastered the art of keeping grass out of the beans," he admits.

Like his neighbors, Moellring does have an occasional insect problem, especially in corn. But he feels the financial and ecological costs of spraying just can't be justified in his operation. "A lot of guys go out looking for bugs so they can spray them," he observes.

Same with fungicides. Moellring never uses them in his wheat; he relies instead on disease-resistant varieties. "If I had a problem, I'd change varieties before I'd use fungicides," he says.

Would he spray if it meant saving a crop from total loss? Probably not, he says. "Even if it wouldn't make a grain crop, it would still make a forage crop," Moellring explains. "I've got livestock. They're a good backup."

Flexible Pasture Plan Anchors His Rotation

The Virginia hill land and river-side bottomland that Sandy Fisher farms could shrug off tens of tons of topsoil in a single hard storm if he'd let it. But with crop rotations and legume-based cover crops, Fisher not only conserves his soil, but has reduced his chemical costs by 25 percent in just one year.

As 1984's Virginia finalist for the National Endowment for Soil and Water Conservation Award, Fisher is gaining recognition for his efforts. But as managing partner for the family-owned Sabot Hill and Brookview farms, with 400 head of beef cattle and 1,300 acres of corn, soybeans, hay, pasture, wheat and barley under his care, he can't lose track of the bottom line, either.

While Fisher has never been a by-the-book chemical user, he's learning to cut back even more. In 1983, he spent $18,000 on chemical fertilizers and pesticides, compared with more than $24,000 the year before. The savings are partly due to a reduction in Fisher's corn acreage, but he's been able to reduce chemical use in all his crops by: replacing purchased nitrogen with legume-fixed nitrogen, no-till planting hay and pasture crops without herbicides, taking a closer look at fertilizer recommendations, and finding ways to reduce herbicides in row crops.

For Fisher, a pasture is much more than a place to turn out the livestock and forget them. It's a highly productive asset, providing most of the beef herd's feed for more than 10 months of the year. He keeps 550 acres of steep hill land in soil-protecting pasture and hay crops. Besides forage, he takes about 250 round bales of hay for his own needs, and sells an additional 25,000 square bales off the farm. Fisher no-tills a variety of crops without herbicides to keep pasture and hayland in high gear.

Sandy Fisher admires alfalfa-orchardgrass pasture that he no-till planted without herbicides.

Field crops are grown on 250 acres of rolling land and 500 acres of flat bottomland adjacent to the James River. The bottomland wouldn't be much of an erosion worry if it didn't flood every year. But in early spring, Fisher's best corn and soybean land is 2 feet or more underwater. Still, this land is tiled, and usually drains quickly to provide a normal-length cropping season.

Rotations, an Austrian winter pea/rye winter cover, and no-till planting with highly controlled herbicide applications also keep this land protected and productive.

Fisher also uses strip-cropping and contour-cropping to help hold soil in place. "Sandy is clearly one of the leaders in our region," says Denise Doetzer, district conservationist for the Soil Conservation Service. "We have field days on his farm to show that what we're talking about can really work. Most of his land has a soil loss of less than 3 to 4 tons per acre."

The centerpiece of Fisher's flexible pasture plan is an 18-row Lilliston 9680 no-till drill. "The Lilliston gives me the ability to plug in crops where and when I want to," he says. And the heavy-duty machine cuts a clean planting slot at the correct depth under virtually any conditions.

One example of the increased productivity made possible by the drill is a winter wheat/Austrian winter pea blend which is no-tilled without herbicides into grazed-down pasture sod anytime during late summer or early fall. "I won't need herbicides there because I'm planting into grass that's been grazed back tight and has just about quit growing by October," Fisher explains. But the wheat/

Reusable Ag-Bag silage bags help Fisher supply his 400 head of Angus-Hereford beef cattle with high-quality winter feed. He ensiles hay taken during wet spring and fall periods.

Lilliston no-till drill is the centerpiece of Fisher's flexible pasture plan. "The Lilliston gives me the ability to plug in crops where and when I want to," he notes.

peas crop, planted at the rate of one bushel of wheat and 15 pounds of peas per acre, does grow quickly enough to provide lush, nutritious forage until the snow flies.

Austrian winter pea seed costs about 70 cents per pound, and Fisher says it produces "just a beautiful crop. They look just like garden peas, and the cows love 'em." He credits the fast-growing peas with fixing at least 100 pounds of nitrogen per acre in his soil.

With 400 head of Hereford-Angus cross beef cattle, and plenty of horse owners in the nearby Sabot and Richmond, Va., areas, Fisher says he'd be "ignoring the signals" if he didn't grow plenty of hay for his own use and for sale. Again, the Lilliston plays an important role. Much of Fisher's 1984 alfalfa/orchardgrass hay was no-tilled into existing sod late in the season, with no herbicides needed. He plants 12.5 pounds each of alfalfa and orchardgrass per acre to establish a new hayfield virtually anywhere he wants. If it's on land that isn't too steep, he can bring it into field crop rotation at any time. And when he does, he'll be able to reduce his nitrogen rate for the field crop by at least 40 pounds per acre the first year, he says.

Fisher usually applies no nitrogen to his hay, relying instead on manure and legumes like alfalfa, red clover and Austrian winter peas. Since beef cattle are on pasture most of the time, their manure represents a substantial cycling of nutrients back on the land, he points out. Keeping the herd on pasture also reduces manure handling chores.

Fisher drags the fields twice a year to spread the manure and get it into the soil faster. This natural cycle is so efficient that, even with 400 head, he has no extra manure available for use on his field crops.

Rounding out the pasture fertility program is diammonium phosphate, applied at the rate of 160 pounds per acre each year.

To describe how he handles his field crop acreage, Fisher uses the example of three 50-acre bottomland fields located in the floodplain of the James River. In 1984, Field 1 was a mixture of SUDEX sudan grass, LAREDO soybeans and black-eyed peas. It was cut three times for hay.

Field 2 was in irrigated corn, which made 135 bushels per acre; and Field 3 was in irrigated soybeans, which made 45 bushels per acre.

In spring '85, Fisher will moldboard-plow the hayfield (one of the rare occasions he uses this implement) or chisel it with his Glencoe chisel plow, then plant corn. The plowing will effectively incorporate the hay residue and help control Johnsongrass, his most serious weed problem. It will also allow him the option of cultivating for weed control after planting. Since the corn will have followed a nitrogen-fixing hay crop, Fisher will cut his nitrogen rate by a conservative 30 to 40 pounds.

The field that was in corn will be in no-till beans. Fisher says he'll apply pre- and post-emergence herbicides sparingly, with small spray nozzles and at high pressure. He uses a soybean oil surfactant to improve their effectiveness and reduce the amount needed.

Finally, the field in soybeans will be planted to hay.

For P and K, Fisher applies diammonium phosphate and muriate of potash as needed. "I don't think of these as 'hot' substances," he says. "I don't think there's any harm done to the soil." He bases his application rates, which are usually lower than those recommended by the lab, mostly on field history and past soil test results. "Conserving soil and being stingy with . . . chemicals is good for the environment, and good business, too," he quips.

While it's obvious that livestock can help reduce the need for many chemicals, a farmer need not be "tied down" with animals year-round. The recent development of portable electric fencing allows farmers to purchase calves in spring, graze them, and sell them for finishing, without devoting the same fields to pasture year after year.

Other possibilities include renting fields in pasture to nearby livestock farmers; or contracting hay-making to a farmer with haying equipment, in exchange for either a reduced price on the entire crop, or simply half the harvest. "That way, you've got 50 percent of it gone right away," says Liebhardt. "Of course, you'd have to have a place to store it; or maybe just sell it right out of the field."

Cover Crops Earn Cash, Improve Soil

Livestock or not, farmers determined to work legume sods into their rotation will often find an angle even where none currently exists. Take Terry Holsapple. He didn't have any particular chemical in mind when he decided to cut costs on his southern Illinois cash-grain farm. His goal was—and still is—to eliminate all of them. "You've got to have a goal out there, and to me, that's got to be it," he asserts.

For easier management, Holsapple trimmed his farm size from 900 acres to 500 in 1984, and put his 120-horsepower FWD White tractor up for sale. "Don't need it," says Holsapple, 31. "I use it for about 100 hours, then it sits around for 362 days."

He also began developing a crop rotation that would not only regenerate his heavily row-cropped soil, but provide a yearly income from all of his fields.

Much of Holsapple's current acreage has been in his family for three generations, and when he took it over in 1974, he had no intention of breaking the corn-soybeans tradition that had endured there for more than 20 years.

Hairy vetch that Holsapple first planted experimentally now helps him reduce N rates and control weeds with fewer chemicals in corn. He combines seed, too.

"For sale, cheap!" Terry Holsapple's FWD White "Field Boss" has outlived its usefulness, now that he's reduced his farm size and is planting fewer row crops.

All that began changing in 1980, when he conducted an experiment on an old, 13-acre alfalfa field. He applied 100 pounds of nitrogen per acre, about 65 percent of his normal rate, then plowed down the sod. "We just whipped her under and planted corn, and got 160 bushels. The point is," he says, pausing for added emphasis, "*we grew 160-bushel corn on that ground, when across the road* (in a corn-soybean field) *we could only get 100, maybe 110 bushels.*"

That same year, Holsapple planted corn on some newly rented land that had been in red clover hay for two years. Again he cut his nitrogen rate by one-third, and again his 150-bushel yield exceeded any he'd ever had. "When those things start happening, you know you should be rotating," he says.

Continued success on the experimental field, plus trial plantings with legumes like hairy vetch and sweet clover, have helped cut his fertilizer bill in half, reduce his herbicide costs to just $5 per acre, and eliminate insecticides on all of his fields in just a few years.

Holsapple's experimental rotation begins with alfalfa, which generally allows four hay cuttings per year in his area. In its second year, Holsapple will mow the alfalfa once, around Memorial Day, then plow it under in July and plant buckwheat. To replace the large amount of potash removed by the alfalfa, he'll apply up to 100 pounds of 0-0-60 per acre.

Holsapple says buckwheat should mature quickly enough to provide a seed crop in the fall. And with an average yield of 30 bushels per acre, he's calculated that it should earn as much as $8-per-bushel soybeans. "The thing is, buckwheat requires no fertilizer, no herbicides, no nothing," he says, pointing out that the potash is actually intended for the following year's corn.

If the buckwheat matures late, he'll follow it with rye. But if he can harvest it early enough, he'll immediately plant hairy vetch for seed. Having experimented with vetch during the federal Payment-In-Kind program in '83, Holsapple is convinced that the crop will help control weeds and boost the fertility of even his poorest ground. In fact, in August '84, four acres of vetch he'd planted a year earlier "on ground that wouldn't grow anything else" yielded 250 pounds of seed per acre. The price of vetch in fall '84 was roughly 50 cents a pound, he says.

Corn will be planted in the third year of the rotation, and Holsapple figures he'll need few, if any, chemicals for a successful crop. "By then, you've dealt your weeds a pretty good blow," he notes.

He'll overseed corn with a legume—probably sweet clover—which will provide yet another seed crop the next year. Then he'll resume the rotation with alfalfa.

Curiously absent from the scheme are soybeans, which in '83 occupied 300 acres of Holsapple's land. "I don't like growing the durn things," he says bluntly. "You fight the grass, you fight the weeds . . . too much risk."

While some of this may sound a bit pie-in-the-sky, Holsapple emphasizes that he's already tried out most of his ideas, and needs only to combine them into one management program. For example, in 1983 and '84, he cut his fertilizer rates by more than 50 percent in fields where he planted corn after hairy vetch or sweet clover, with no difference in yields. "The (corn's) color and height

Holsapple checks sweet clover seed in early July, a few weeks before combining. Sweet clover is one of three income-producing cover crops in his experimental rotation.

are so much better, and the ground's a lot looser," he says. "I'm pretty sure it'll work. It's just getting the stands I need."

Banding rather than broadcasting fertilizers has been a big help, too, he adds. "The universities say you can cut your rates in half if you do that."

Though Holsapple now buys less than half the total N that he used to, replacing anhydrous ammonia with a more expensive urea-ammonium nitrate mix has kept his nitrogen bill about the same. But he's reduced his potash rate from 60 pounds per acre to just 30, and has all but quit buying phosphorus fertilizer. "There are millions of pounds of phosphate down there that corn and soybeans can't use but alfalfa can," he says. He now spends just $30 per acre for fertilizers, half what he was spending three years ago.

Finally, on fields where row crops have followed sweet clover or vetch, Holsapple has noticed a decline in weed populations. He uses about $5 worth of herbicides per acre, and relies on his six-row Noble cultivator with Danish tines to clean up the survivors. "Once you've had your ground in alfalfa for several years, you have pretty good weed and grass control," he explains.

Insecticides? Holsapple doesn't use them, and says he probably never will. "Just because you have a problem this year, that doesn't mean you're going to have one next year," he says. "But if you spray, you're going to kill every single beneficial insect there.

"The rotation's the thing," he adds. "If you rotate, you won't have any problem. I just try to relate the crop that I'm going to be growing next year to the one I'm growing this year."

Rotation Replaces Insecticides

Maurice Jackson has literally made a science of reducing insecticides on his 1,300-acre, cash-grain/livestock farm in northwestern Illinois. Though he seldom used them on his 300 to 400 acres of hay and pasture, Jackson was spending more than $8,000 a year on insecticides for his 900 acres of corn. Due largely to his own on-farm experiments, he spent less than half that in 1984.

"With observation the year prior to withdrawing chemicals, I questioned whether we really needed them," says Jackson, 54, who raises 175 Simmental beef cows and feeds about 250 Simmental calves. "I just did without them on a few acres. The only way you can tell if you can do without them is to observe.

"That (insecticides) appeared to be an area where I felt a dollar invested wasn't a dollar returned," he adds. And he may be more right than he thinks. A USDA study titled "Returns to Corn Pest Management Practices" shows that the estimated return for every $1 farmers spent in 1983 on insecticides and herbicides was $1.03 and $1.05, respectively. And that doesn't include application costs. (Single copies of the study, #AER-501, are available for $2.50 from: Superintendent of Documents, U.S. Government Printing Office, Washington, D.C. 20402.)

Actually, Jackson did without insecticides on more than just a few acres in '84. He eliminated them completely on 300 acres of corn that had followed alfalfa, and reduced his rate by 40 percent on nearly all of his remaining corn acreage, saving $4,560 in the process. In one 80-acre test field, Jackson compared Dyfonate-treated corn with untreated plants in side-by-side, 12-row sections. The treated rows received 4 pounds of Dyfonate per acre, about 60 percent of his normal rate. By early July '84, when rootworm damage should have peaked, both plots were growing vigorously. "That was enough to convince me to take a closer look at insecticides," says Jackson. "That's why you're seeing 300 acres without them."

Eliminating herbicides has been more difficult. Jackson tried controlling weeds without chemicals in a 40-acre cornfield, but learned quickly that, without an established rotation, at least some herbicides would be necessary for awhile. "I hoed that three times and cultivated twice," he recalls. "We did make it work, but I sure wouldn't want to do the whole farm like that. That (rotary) hoe doesn't do a perfect job. Otherwise, I'd rely on it 100 percent."

Instead, Jackson now rotary-hoes his corn when the plants are about 4 inches tall, and applies Banvel for broadleaf control in the same operation. Then he cultivates once or twice. The combination enables him to get by with "a minimum rate" of Dual for broadleaves, and just 75 cents worth of Banvel per acre.

Jackson is planning a long-term strategy for reducing fertilizer costs, too. His first step was to choose a soil testing lab that was a bit less "generous" with its fertilizer recommendations. "When we got to looking at our soil tests, we found we were over on nitrogen," he notes. To test his theory, Jackson reduced his nitrogen rate by 40 pounds per acre on the 300 acres of corn that had followed alfalfa. "At the moment, it looks good," he said in spring '84. "Compared to what we had in the past, we've cut back quite a bit (on nitrogen fertilizer).

A banding kit mounted on his rotary hoe enables Jackson to get by with just 75 cents worth of Banvel per acre in corn.

"We're probably thinking that 130 or 140 units (pounds) would be our *maximum*, now," adds Jackson, pointing out that 150 to 200 pounds was a normal rate just a year or two earlier. "About 70 units would be our minimum."

Jackson applies one-third of his nitrogen at planting, and the rest at second cultivation. "The later nitrogen application ties right in with your cultivation," and helps aerate the soil for better root structure, he explains.

He has also replaced acidulated phosphorus fertilizers with rock phosphate, and begun using 0-0-50 (sulfate of potash) or K-Lime (a cement industry by-product that contains calcium, potassium and sulfur) instead of 0-0-60 to supply potassium.

Jackson is convinced that any attempt to reduce chemical use must begin with a balanced approach to soil fertility. "You've got to know where your soil is before you know where you're going," he emphasizes. "I have a feeling—and this was told to me by a soil consultant—that the more balanced my soil is, the less I'll need insecticides."

In fact, a good soil fertility program, which includes soil testing *and* plant tissue analysis, can help farmers save money on fertilizers, too. For example, an 18-year study at the University of Delaware showed that phosphorus and potassium remained very high in soil that received no P or K fertilizer for nearly two decades (see "18 Years Of Top Yields—Without P and K," *The New Farm,* March/April '84). Soil tests taken at the beginning of the study showed high and very high levels of P and K. Yet despite corn yields of up to 240 bushels per acre, there was no significant yield difference between plots that received as much as 50 pounds of P fertilizer and 100 pounds of K fertilizer, and those that received none at all. For a farm with 500 acres of corn, the total cost of unneeded fertilizer for the entire study period would have been $202,950.

A fluke? Not at all, says RRC's Liebhardt. In soils with a history of high P and K levels, studies show that neither yields nor fertility will suffer when phosphorus and potash are withheld. "The Tennessee Valley Authority and land grant universities have 22 experiments in 10 states . . . and they have shown that corn yields do not drop if you reduce P or K, or even if you eliminate it."

Few farmers know that better than Dorsey Owings, who raises corn, soybeans, vegetables and hogs on 3,500 acres in Maryland. After several years of applying "maintenance" P and K doses to soils that already had plenty, Owings eliminated K and reduced P by two-thirds on his 1981 sweet corn. His crop did fine, and he saved $60,000 (see "They Saved $60,000 On P and K," *The New Farm,* February '83).

Owings did the same thing in '82, and saved even more money. "I haven't bought any potash for two or three years," he says. "My P and K levels have been high to very high. If they start to fall off, I may apply more. I'm just not going to put any more on until I need it."

Just how long can high-phosphorus soils go without additional fertilization? Auburn University researchers set out to answer that question by halting P applications on some fields in 1929. "And they still don't require any," says Dr. J. T. Cope, an Auburn soils professor. What's more, scientists at Auburn and the Tennessee Valley Authority now say that up to half the farmland in 25 states has high P and K levels from years of fertilizer applications. "It's obvious we are getting a build-up," observes Cope.

That doesn't mean all farmers should quit applying P and K, Liebhardt quickly notes. But it does mean they should seek advice from land grant soil fertility specialists if they suspect they could get by with less. Meantime, he adds, plant tissue analysis can provide a more complete view of a soil's ability to support good yields. In fact, it was

Maurice Jackson checks Dyfonate-treated corn for rootworm damage in midsummer. Untreated corn had no higher rootworm populations than corn receiving 4 pounds of Dyfonate (about $6 worth) per acre.

Tissue Testing Can Save You Money

His soil tests showed a low level of phosphorus in a 200-acre cornfield, so the farmer applied 50 pounds of phosphorus fertilizer per acre like his lab advised. But this year, he also left a check strip with no P fertilizer.

Tissue tests showed no difference in phosphorus uptake between the fertilized and unfertilized plants, which meant there was more phosphorus available than the soil test actually showed. But where?

The farmer decided to find out. He tested his soil at both 6 inches and 18 inches deep the following year, and soon learned that a large amount of phosphorus—more than enough to meet his corn's needs—was already present in the lower level. He applied no P fertilizer that year, and saved $2,500.

Sound farfetched? It's not. In fact, scientists and farmers are becoming increasingly aware that both careful soil testing and plant tissue analysis offer many cost-cutting advantages over routine soil testing alone. For one thing, tissue tests reveal nutrient levels from a much larger area of a field—the plant's root zone—than core samples. They also give a more accurate reading of micronutrients, and show how a plant's nutrient uptake is being affected by tillage methods, changing weather patterns, and other factors. "In a sense, they let the plant tell the farmer how it feels," says Dr. William C. Liebhardt of the Rodale Research Center.

That doesn't mean soil tests should be eliminated, Liebhardt quickly notes. They're still a good guide to soil nutrient levels. But combining them with plant tissue analysis enables a farmer to develop customized fertilizer application rates based on data from his own farm.

Sampling the right part of the plant at the proper growth stage is critical. For corn, tissue samples should be taken from the ear leaf at silking. Soybean samples are taken from a fully expanded trifoliate leaf near the top of the plant at early bloom. And for small grains, select a flag leaf (the uppermost leaf near the seed head) at or just before heading.

A good corn or soybean sample consists of 30 to 50 leaves selected randomly from a field. Small grain leaves aren't as large, so you'll need more.

Place the leaves in a bag (cut them up if they won't fit) made of anything but plastic, so air can circulate. Then, mark the bag to identify the field, and take the sample to a lab within six to eight hours for drying. If necessary, tissue samples can be dried overnight in an oven heated to 140-150 F.

Be sure to include a note telling the lab what nutrients you want included in the analysis. Liebhardt suggests phoning the lab first, to find out what tests are offered and how much they'll cost. For about $10 to $20, most labs will test for all major and minor nutrients except sulfur, which is optional in many areas. Labs may also offer guidelines for taking the samples, and provide bags for shipping.

When the analysis is complete, you'll receive a printout similar to a standard soil analysis. Compare the results with soil tests and fertilizer applications in the same field—just like the farmer in our example did—and you'll soon be making more informed, cost-effective decisions about fertilizer.

---

plant tissue tests that convinced Dorsey Owings that his reduced 1981 sweet corn yields weren't due to a lack of P and K. Yields had fallen to 4.5 tons per acre, but Owings felt that heavy weed pressure—the result of rain washing away his herbicide—was the culprit.

Tissue analysis confirmed his suspicion: Both P and K levels were "high." "I'll probably continue to do tissue tests from now on," he says. "They give me a good spot check on the nutrients available to the following crop."

Landlords Love This Cost-Cutting Plan

Some landlords and bankers feel that a reduction in chemical use automatically means a reduction in yields and cash flow. Likewise, highly leveraged farmers, or those who rent a good portion of their land, may be uneasy about management changes that focus on more than just short-term profits.

In such cases, farmers might try out their cost-cutting ideas on a few home acres, first. Then, keep accurate yield and cost figures to show how much money and soil can be saved if the new methods were used on the whole farm.

Farmers with little or no home acreage may want to concentrate on techniques that can show immediate results to their landlords in the way of improved erosion control, lower production costs, or both.

One farmer who's done that quite successfully for several years is Donn Klor, who leases his entire 500 acres in central Illinois from five different landlords. About half of Klor's land is sloping, and about one-fourth is "timber soil" made up of silty clay-loams, which his county soil map describes as unsuitable for row-cropping with conventional tillage methods. Yet Klor's landlords, who share one-half the cost and income from the land, expect him to maintain a corn-soybeans rotation. "A lot of things go unsaid," says Klor, 37. "But you know them and you know what they'll let you do. They don't want stock. They're used to the way things have been for the past 20 years."

No-till seemed like a reasonable alternative, he says, "but then you'd have to load back up on the chemicals, which is what we're trying to eliminate in the first place.

"We have no-tilled in the past on a limited basis, and found that we couldn't afford all the ' savings,'" he adds. "After applying $60 worth of herbicides and then having to use a special residue cultivator to salvage the crop for two years in a row . . . (that) doesn't put any money in your pocket or raise your enthusiasm much."

In 1983, Klor decided on ridge-planting. Since then, he's cut his fuel and herbicide costs by more than 50 percent in his 275 acres of soybeans and 225 acres of corn, and reduced soil erosion to less than 6 tons per acre annually on even his hilliest ground. How do his landlords feel? "I'm blessed with cooperating landlords who can see the benefits," Klor says. "Their farms will benefit from conservation, and, by using less herbicides, they will net more profit for themselves."

As the first farmer in Sangamon County, Ill., to try ridge-planting on a large scale, Klor was forced to buy new equipment. That required some cooperation from a Production Credit Association loan officer and a local equipment dealer, who combined to offer Klor an attractive equipment trade and lease-buy agreement on a Hiniker Econ-O-Till cultivator and a Hiniker ridge-planting attachment for his John Deere planter.

From there, Klor had to decide which chemical to reduce first. Cutting back on fertilizer might have meant lower yields, so he chose herbicides. "I'd done everything else I could do," he recalls. "I'd changed my tillage. Herbicides seemed like the next thing to work on."

This ridge-planted corn received just $12 worth of herbicides per acre, less than half Klor's normal rate. "You don't see any weeds in there," he says.

Klor shows off $150 banding kit that paid for itself after he'd sprayed just 10 acres. He cut his herbicide bill in half by banding.

Shields on Donn Klor's Hiniker Econ-O-Till cultivator protect small plants during cultivation. Disk hillers pull weeds away from row; coulter and sweep destroy them. V-shaped ridging wings atop sweep shank are lowered into place at last cultivation.

But he soon learned that relying completely on cultivation would not provide adequate weed control. The farm's size made timely cultivations by just one or two people nearly impossible. So he bought a six-row banding kit for $150, and began applying herbicides only where necessary, and at 50 percent of his normal rates. The idea worked so well that the banding kit paid for itself in herbicide savings after Klor had used it on just one 10-acre field. "If you can cut your costs in half by banding, then eliminate half your banding, you've made a significant reduction," says Klor.

In fact, he adds, he could probably eliminate grass herbicides in corn if he had a bit less acreage to cultivate. He plans to replace herbicides with his rotary hoe on a few acres of beans, too. "The weed pattern changes in four to five years with ridge-tillage, so I'll just have to experiment and wait and see," he notes.

Klor uses pheromone traps, which he gets free from his county Extension service, to monitor cutworm populations in his corn. Though he had used Counter, Lorsban and even Aldrin to control cutworms in the past, he's found little use for them, lately. "As long as you rotate corn and beans . . . the likelihood (of insect problems) is much removed," he explains. "This year, we just didn't have any amount to speak of. It just doesn't warrant putting $10 or $12 out there to kill a bug you're not going to have."

With bugs and weeds under control, Klor now has set his sights on reducing his $90-per-acre fertilizer costs. On one of his cornfields, he tried injecting 28-percent nitrogen fertilizer instead of broadcasting it, hoping to replace the amount of nitrogen lost to the atmosphere. "Theoretically, we should be able to either apply the same amount and get a better yield, or apply less and get the same yield. That's what we want to find out."

He's also replacing his 180-60-80 fertilizer with dry forms of P and K, and will apply these nutrients at reduced rates on some of his fields. "We're going to be a lot more critical with our soil testing," he notes. "When a lab tells you you don't need any more P and K, and the dealer says you need more for maintenance . . . I think we're going to try not putting any on, and see how it works. That's $40 we don't need to spend."

Diversity, Marketing Mean Profits

Often, a farmer will give up some or all of his rented land, and raise higher-value crops on his home acreage to make up for the lost income. That's exactly the decision Frank Stancil made in the late '70s.

The way Stancil sees it, farmers like to work the land for a living because they need a certain amount of independence. Never mind the hard work, "just let me be my own boss," he says.

So it was with some dismay that Stancil saw his management alternatives becoming fewer and fewer during the late '70s. Because of rising production costs and poor crop prices, he reached a point where he thought he'd have to expand fast to "lower his per-unit cost" as the experts advised, or give up his Tennessee farm. "In 1979, I spent $13,500 on chemicals, but I still lost money," says Stancil, who was raising mostly soybeans on his own 100 acres and 400 rented acres.

Five years later, in 1984, he spent only $600 on synthetic fertilizers and pesticides, and earned higher profits by raising more than a dozen different crops on just 100 acres. "I'll never go back to having just a few options," says Stancil. "To stay in it, I would have had to buy a bigger tractor, planter, and a combine. That's a lot of high-priced equipment to have just sitting around looking pretty for eight months of the year."

Now he has more control over his costs, and, just as importantly, over the prices he receives. For example, Stancil still grows soybeans, but only on 40 acres. The crop is grown without chemicals, and sold at a premium price to a nearby manufacturer of soymilk, tofu, and soymilk "ice cream."

He's still growing corn, too. Popcorn, that is. And he sells it at above-wholesale prices to regional retailers.

Finally, Stancil has begun growing a number of vegetable crops. But instead of selling them to a wholesaler, he takes them to a Nashville farmers' market or delivers them to customers who are happy to pay supermarket prices for farm-fresh produce brought to their door. "I get $17 per bushel for green beans when I sell them by the pound, compared with $6 per bushel wholesale.

"We farmers have to pay more attention to marketing," Stancil adds. "Where would other industries be today if they just pushed their products out the back door and said 'I'll take whatever you'll give me for them.'"

A lot of things happened at once to make 1980 a turning-point year for Stancil. Of course, there was the beating he took on his soybean crop the previous growing season. But there was an educational process going on, too, which included contact with other members of the Tennessee Alternative Growers' Association, a visit to the Rodale Research Center, and a talk about the practical side of regenerative farming with Ben Brubaker, who farms about 320 acres adjacent to the Rodale Research Center (see "Good Yields For 14% Less," *The New Farm*, Nov./Dec. '83).

That year, Stancil dropped all but 180 of his rented acres, and began using fewer chemicals on his corn and soybeans. He withheld fertilizer and herbicides from 15 acres of corn following a hairy vetch-rye cover, "just to see how it would do." In another field, more than 30 acres of soybeans were planted after corn, again without fertilizer or herbicides. The corn yielded 100 bushels, and the beans 35; good for dryland crops in his area, and right up with his other, more conventionally treated fields. Stancil knew he couldn't maintain that performance, though, because he didn't have enough cover crop or hay acreage to plant row crops after legumes every year.

But the evidence was there: He could get good yields with more reliance on crop rotation and cultivation, and less on chemical fertilizers and pesticides. He knew that a legume crop, which could be used in the rotation for plowdown, cover, hay or seed, was needed to fix nitrogen and build biomass quickly without shutting off cash flow.

The rye-vetch mix turned out to be the answer. In one 12-acre field, Stancil planted corn in '81, followed by full-season soybeans in '82. After harvesting the beans, he planted the rye-vetch mix at the rate of one bushel of rye and 20 pounds of vetch per acre. The mix germinated, overwintered, and regrew vigorously during 1983.

Some might say Stancil made too big of a sacrifice in the name of reducing chemical costs on that field. After all, wasn't the field idle in '83? Didn't the lack of a cash crop hurt his bottom line?

"No way," Stancil replies. "It'd be conservative to say that the vetch fixed 150 pounds of nitrogen per acre." But that's not all. Stancil says the vigorous vetch-rye growth outcompeted weeds all season long, to set the stage for good non-chemical weed control the following season. The cover crop also produced tons of biomass to build soil tilth and improve water-retention, important factors anywhere, but especially in the South.

Finally, Stancil combined enough rye seed for his own use, plus 400 pounds of vetch seed per acre from the cover crop, to be sold for 35 cents per pound. These were cleaned with an old Clipper seed cleaner and separators that Stancil also uses for custom work.

All that for only a nominal cost to establish the crop. "Not bad for 'idle' land," he suggests.

Along with Stancil's other changes, livestock have taken on a new importance. He has long raised about 20 head of beef cattle in partnership with his father, but now he is planning his manure applications to make sure they end up on fields where nutrient needs will be highest. Now that he has more hay and pasturing possibilities in his rotation, he plans to steadily increase the number of beef cattle on the farm.

Stancil's current fertility program includes green manures, livestock manure, rock phosphate, and the wetting agent Basic H. The major active ingredients in Basic H are surfactants, chemicals commonly found in soaps, detergents and emulsifiers. Such materials are said to "make water wetter," thereby improving its ability to soak into soil.

Some wetting agent manufacturers have claimed their products unlock nutrients in the soil, too. Stancil's not sure about that, but he does feel that the up to $10 worth of Basic H per acre that he applies has improved soil drainage and water-holding capacity.

Another use for Basic H is as a surfactant blended with herbicides. Stancil says he can cut his herbicide rate in half and still obtain the same results when he mixes in Basic H. For more information on Basic H, see "The Thompsons Test Basic H," *The New Farm*, March/April

Hay is becoming more important on Frank Stancil's farm. He uses legumes for livestock feed, seed crops and nitrogen.

'83; or the book "A Practical Guide To Novel Soil Amendments," which is available from the Regenerative Agriculture Association.

Stancil applies rock phosphate at the rate of 1,000 pounds per acre when soil tests show declining P levels.

Stancil has depended on his rotations and cultivation to reduce weed pressure to a certain extent, but banding a post-emergence herbicide into the row is still an occasional part of his program. He uses a rotary hoe for early weed control, and a four-row cultivator fitted with either spring teeth or sweeps for later work.

More Profit With Less Corn

Dairyman Bob Fogg had always used a planned crop rotation on his 260-acre Michigan farm. So, for him, cutting chemical costs began with reducing his corn acreage by more than 50 percent in 1981, and each year devoting more land to alfalfa. "By gradually raising less and less corn, you always have a good sod to plow down for the corn you do raise," he notes. "I just like to rotate well enough to eliminate all of my insecticides and purchased nitrogen."

Fogg generally applies no purchased nitrogen to corn following alfalfa, and about 60 pounds per acre to corn following clover. Manure from his 40 cows and 50 heifers is spread two or three times a week during spring and summer, and kept covered and tightly packed during winter storage to minimize nutrient loss.

But that still hasn't enabled him to quit buying P and K fertilizers, "mainly because I raise so much alfalfa, and I have to use so much more phosphorus and potassium (than he gets from the manure).

"I still use phosphorus (0-45-0) and potassium (0-0-60) on the hay, and my oats get a little starter," Fogg adds. "You've got to get P and K from some place, and a certain amount of commercial fertilizer is still an economical source for me. Those are probably the last two chemicals I'd eliminate altogether."

That's not to say he isn't trying. In 1981, Fogg set aside an 18-acre field that had been in alfalfa for several years, and began experimenting with ways to cut costs even further. He plowed down the alfalfa, and planted corn without synthetic fertilizer or manure. For weed control, he sprayed 1.5 pounds of atrazine per acre, and cultivated once or twice. "That was the best corn I've ever raised," Fogg says of the 90-plus bushels per acre he harvested from the experimental field. "It was beautiful."

The following year, Fogg planted corn again in the field, this time with no chemicals whatsoever. He fertilized with 10 tons of dairy manure per acre, and cultivated twice. Yields fell to about 80 bushels per acre, and, not surprisingly, quackgrass, velvetleaf and redroot pigweed were much more of a problem. "It wasn't quite as good as the year before, but it was by far cheaper to grow," he quickly emphasizes. "When you do it that way, you feel like you're making more money, because you don't have to pay the bill at the co-op."

Fogg cut the corn for silage that year, and sowed rye in fall, again without fertilizers or pesticides. The crop yielded 25 bushels per acre, and was virtually weed-free. "Rye's never weedy," he says. "Seems like it's 2 feet tall before any weeds can even start growing."

After harvesting the rye in '83, Fogg planted clover. He plowed that down in '84, spread 10 tons of dairy manure, and planted corn with no purchased nitrogen (instead of the 60 pounds he normally applied to clover). Unfortunately, the 2-year-old seed he'd used germinated poorly, and Fogg had to spray some 2,4-D just to keep weeds from overrunning the field.

But 1984 wasn't without its successes. On an additional 24 acres of corn, which had also followed alfalfa and received 10 tons of manure per acre, Fogg decided to use up a few hundred pounds of 16-16-16 starter fertilizer left over from one of his oat fields. He applied 100 pounds per acre, but the material ran out after just seven acres, and Fogg had to rely completely on the manure and legume to fertilize the other 17 acres. "You can't tell where I ran out of fertilizer," he says. "The grass and weeds seem to be growing just as well on either side, too."

Fogg says he now spends just $30 an acre to grow corn (including seed and fuel), and has cut his overall chemical bill by $35 per acre since 1981. Growing less corn—40 acres instead of his usual 100—has been a big help. But refusing to spend money on any product he feels won't pay for itself is the biggest change he's made. Take nitrogen fertilizer, for example. "That's probably the most expensive item there is, and it's going to be more so in a few years," he predicts. "Yet it's one of the easiest items for a farmer to avoid buying. I mean . . . he can grow his own!"

He feels the same about insecticides. His crop rotation helps keep bugs and weeds from getting out of control, and lets him reserve chemical warfare for emergencies. In

Bob Fogg inspects rolling shields on cultivator. S-shaped Danish tines throw dirt beneath crop leaves for in-row weed control.

'WHERE DO I START?'

Rotations Worth $1 Billion... But Why?

If the corn and soybeans that U.S. farmers now plant continuously were grown in rotation instead, these crops would earn an extra $1 billion annually, and become 50 percent more profitable.

Why? According to Extension Agronomist Kent Crookston of the University of Minnesota, it's simple: Crop rotations increase yields. Period.

Never mind pumping more irrigation water, Crookston writes in the March 1984 issue of Crops and Soils Magazine. Never mind heavy applications of fertilizers and pesticides. "Research trials have established that yield increases from rotation persist beyond optimum levels of fertility, soil tilth, soil moisture and pest control. Some unknown factor (or factors) result in a rotation yield benefit which has not been adequately explained." In other words, whether you use chemicals heavily, occasionally or not at all, crop rotations can improve both your productivity and profit.

And Crookston isn't talking about tiny improvements, either. He cites one study from the central Corn Belt in which corn rotated with soybeans yielded 13 bushels more than continuous corn, and earned $41 per acre more profit.

A six-year study by the University of Nebraska offers similar results: Yields from continuous corn were as much as 30 percent below those of corn in a corn-soybeans-corn-oats/clover rotation. And long-term experiments in Minnesota and elsewhere show at least a 10-percent yield increase in both corn and soybeans when these crops are rotated.

In fact, given 1983's average corn yields and production costs, a five-bushel yield increase can boost net income by 50 percent, Crookston says.

Pinpointing the biological reasons behind the "rotation effect" hasn't been easy. But scientists agree that the economic advantages are clear: Rotated crops simply cost less to grow; about $15 per acre less in the central Corn Belt study, and as much as $31 per acre less in the Nebraska experiment. Lower fertilizer and pesticide costs were cited in both cases. "Where cash flow is a concern, systems with lower cash outlays may have advantages, apart from the profitability issue," say U of N agricultural economists. "Clearly the major difference in cash outlay is between continuous corn and the rotational systems."

Are these benefits limited to corn and soybeans? Not at all, says Crookston. "Many crops, such as wheat, sunflower, sugarbeets, cotton, sorghum and barley reportedly benefit from the presence of a different crop the preceding year.

"It should be emphasized," he adds, "that even though the rotation effect cannot yet be satisfactorily explained by scientists, it can be exploited by farmers–every year."

---

fact, Fogg hasn't used insecticides for eight years, and isn't sure he ever will again. "It would depend on how important the crop was," he says. "If it was five acres of oats, I don't think I'd worry about it."

Because he feeds nearly all of what he grows, Fogg may worry less about crop damage than a cash-grain farmer would. But that doesn't mean he's willing to settle for low yields or poor crop quality. To the contrary, his 90-bushel corn, 70-bushel oats and 5-ton alfalfa help him maintain a herd average of nearly 18,000 pounds of milk per year, some 2,000 pounds higher than the county average. Fogg's butterfat content is often as high as 3.8 percent, and he buys *no* protein supplements.

"Most people are afraid that if they don't spend all that money, they'll have a crop failure and they won't be able to make their payments," observes Fogg. "Well, if they don't spend the money, they won't *have* the payments."

So... Where DO You Start?

In many respects, the farmers in this chapter are no different from their 2.3 million contemporaries.
- They're all interested in good yields and crop quality.
- They all want to improve their soil, and safeguard their health and environment.
- And most importantly, they all want to stay in business and remain profitable.

What makes them unique is that, for the first time ever, they're beginning to place *equal* emphasis on all three of these important objectives. For some, that means using a new crop rotation. Others are simply scaling down their farms to more manageable size. New crops, marketing methods or tillage techniques are the answer for still others.

But without exception, these farmers *started* with a single purpose in mind: to avoid spending money they didn't need to spend. They examined their farms closely; determined which, if any, inputs weren't paying their way; then set out to replace them with management methods that were more consistent with all of their objectives, not just one or two.

There's one other characteristic that sets them apart: They're all *determined* to change the way they farm. "There are only so many options," says RRC's Liebhardt. "You can't help a person who won't change his situation."

Fogg points to corn that received 100 pounds of 16-16-16 fertilizer in 1984. Corn at right received none. "You can't tell where I ran out," he quips.

What The Research Reports Haven't Told You

Reports from the Rodale Research Center (RRC) have long been a valued source of information for farmers interested in using fewer chemicals. Frequently, though, some of the best advice on non-chemical production methods is left out of these reports by the scientists who produce them. Why? Well, for starters, such advice may be scientifically "invalid" if it wasn't part of an experiment's original design. Or, the advice may be nothing more than a hunch, an informal inference based on what the scientists *think* an experiment's results mean.

But hunches and surprise results can be valuable, too. At worst, they can lead to a new set of "what ifs," and a new series of potentially revealing on-farm experiments. At best, they can be real moneysavers for farmers.

Either way, they're too useful to be left unsaid. So, in spring '84, *The New Farm* sent Senior Editor Mike Brusko to the RRC with an armful of questions for three scientists who have spent the last several years researching ways to help farmers cut back on chemical costs: Dr. Richard R. Harwood, director of the RRC for eight years; Dr. William C. Liebhardt, RRC assistant director; and Martin Culik, the agronomist who coordinated RRC's five-year "Conversion Project."

Here is the transcript from that meeting.

THE NEW FARM: **The one question that everybody has been asked is: Should my goal as a farmer be to eliminate all use of fertilizers and pesticides?**
HARWOOD: Should we not put nutrients into a system? We'd have to say no. We'd have to put nutrients in. Hopefully, we'd have some decent-cost nutrients that aren't going to be disruptive, so that you don't have to put in an acid or salt and screw up your system in order to get the nutrient in there.

"Good soil nutrient levels are probably more important on a low-input farm."

THE NEW FARM: **OK, but should I be totally self-sufficient for my crop nutrient needs?**
CULIK: Well, I would say as much as possible. But there's a limit in each system: whether you have livestock, whether you have hay in the rotation, the climate.
HARWOOD: You can answer that by saying you get the nutrients only to that level where you're not limiting crop yield, and then you don't go any higher than that.
LIEBHARDT: To farm with nutrients limiting is economically the wrong way to go. I think good soil nutrient levels are probably more important on a low-input farm than on a high-input one. Take weed control. If you have good fertility and a good stand of hay or small grains—even soybeans—the crop canopy will block out light and help prevent weeds. In a conventional system, you can just use herbicides. But a fully closed crop canopy is the best herbicide there is.
CULIK: If I have a large acreage and I can't get out and cultivate weeds, I'm not going to let them go. I have to use something.
HARWOOD: For field crops, we're going to have to argue that, on nitrogen, probably you can be self-sufficient. And most field crop farmers are going to have to be.
THE NEW FARM: **Define field crop.**
LIEBHARDT: Well, those would be different from horticultural, vegetable or fruit crops. In other words: corn, soybeans, hay, grains. Anything other than fruits and vegetables.

"You need to look at what the limiting factor is, and whether you can eliminate insecticides right off, or cut your herbicide use to a fraction."

THE NEW FARM: **And you feel you can be self-sufficient for nitrogen in those?**
LIEBHARDT: Largely self-sufficient.
HARWOOD: A lot of people are. Even in the cash-grain crops, people are doing it.
CULIK: You have to have hay in the system. Or, you can develop overseeding technology.

Reduce First, Quit Later

THE NEW FARM: **What's the most important thing for me to consider during my first year of reduced chemical use?**
CULIK: Don't do it all at once. Don't try to convert your whole farm, or don't eliminate chemicals all at once.

Agronomy Coordinator Martin Culik discusses "Conversion Project" with farmers attending annual RRC field day. Results show that soybeans or small grain-legume hay can be grown profitably without chemicals in just one year.

LIEBHARDT: If a farmer has 200 acres, he should experiment with, say, 20 acres the first year—one field. And not go into "cold-turkey" corn.
HARWOOD: For us, the limiting factor in conversion was clearly nitrogen and weeds. Some of the farmers in your conversion survey (see *The New Farm,* March/April, July/August and Nov./Dec. '84) said that wasn't their problem. So you need to look at what the limiting factor is, and whether you can eliminate insecticides right off, or cut your herbicide to a fraction. Get your rotation started, and then come off in the right place in the rotation. For example, eliminating corn rootworm insecticides can be as simple as rotating to another crop. If warm-season weeds are the problem, switching to something that comes up early—a small grain or hay—can help reduce herbicides.
CULIK: It has to be a planned process.
HARWOOD: Don't make it a Damascus Road conversion with a bolt of lightning.

"A lot of farmers are not going to eliminate everything in three years."

LIEBHARDT: I was just talking to Diane (RRC Entomologist Diane Matthews). She's renting from a nearby farmer, and she said she's been talking to him about using fewer herbicides, so he bought a cultivator. It's the first time he's ever had a cultivator, and he's banded herbicides this year for the first time. There's an example of a guy who's doing it slowly.
THE NEW FARM: **In other words, one of the things farmers should consider is whether or not it's a realistic goal to try and eliminate a whole family of chemicals—whether it's nitrogen fertilizer, P, K, or pesticides.**
LIEBHARDT: It probably is not a realistic goal. A lot of farmers are not going to eliminate everything in three years.
HARWOOD: The first objective has to be to stay in business. And then if you have to buy nitrogen, get away from anhydrous and shift over to urea. A lot of people will use urea because they say it's less disruptive to soil microbes and insects—to soil life. We don't have any data on that, but . . .
LIEBHARDT: I suspect that, in the first year we planted organic corn in our conversion experiment out here, had we banded 6 to 8 inches of herbicide over the row and used 50 pounds of urea, we would have gotten darn near the 125-bushel corn that Penn State says you'll get by following its 150-pound nitrogen recommendation. We'd have gotten very close to it, because we couldn't control weeds in the row. It was the foxtail in the row that killed us, and the lack of nitrogen. And they work together to make the weed problem unbearable.
HARWOOD: And if we'd banded urea, too, it probably wouldn't have been completely disruptive.
LIEBHARDT: The other thing, from the standpoint of nutrients, is that you get most of your response to the first bit you put on. What does Penn State recommend for corn, 130 pounds of nitrogen the first year? Probably 80 to 90 percent of the response comes from the first 50 pounds.
THE NEW FARM: **What about pesticides? Should it be my goal to completely eliminate the use of herbicides and insecticides?**
LIEBHARDT: Instead of approaching it from that standpoint, why not use them only when necessary? If a farmer finds he can't get weed control without some herbicide banded down the row, then maybe that's the way he ought to go. He ought to do it so he can survive economically.
HARWOOD: So the short-term strategy is to minimize purchased inputs, and the long-term is to eliminate them. If you look at herbicide use today, $2 billion worth of

herbicides are used on corn and soybeans, $200 million on all the rest of the crops. There are 10 times more herbicides used on corn and soybeans than on everything else put together. I guess that's because corn and soybeans are a monoculture.

CULIK: Other pesticides can be almost totally eliminated —insecticides, fungicides—because of the rotation and the integration.

HARWOOD: You'll have some things, like potatoes, for instance, that are tough, at least in the East.

CULIK: Or in the Southeast, there may be more need for insecticides on some crops. But rotation will take care of many of these problems.

Legumes Replace Manure

THE NEW FARM: **I don't raise livestock. How can I still become self-sufficient for crop nutrients?**

HARWOOD: The cash-grain farmers who are doing it successfully usually have at least one of those cash-grain crops as a legume. Dried peas, split peas, beans—some kind of grain legume. Then they do some overseeding, or they'll put in some fall and winter cover crops.

LIEBHARDT: In a typical Midwestern corn-soybean rotation, overseeding vetch, particularly in soybeans, would help provide nitrogen for the corn and reduce erosion. You could actually overseed into both corn and soybeans.

HARWOOD: An example is the cash-corn farmers in Delaware. They don't have livestock. They wait till the corn is harvested, then they seed vetch or Austrian winter peas early in September. It establishes early in fall, overwinters, and produces lots of nitrogen-rich biomass quickly in spring. They figure that, for every 12 inches of growth, they get 100 pounds of nitrogen. They're producing the same amount of corn that they would with 100 pounds of nitrogen.

Hay Not Necessary

THE NEW FARM: **Again, what if I don't have the machinery for hay, or buildings for livestock? Do you feel that overseeding is the way for me to go?**

CULIK: Overseeding or green-manure crops. Overseeding techniques are still being researched, but hairy vetch and/or red clover seem to work best, so far. In Pennsylvania, annual ryegrass works well, too. Green-manure crops include rye or barley planted after soybean harvest.

THE NEW FARM: **What can they do for my soil that animal manure would do?**

LIEBHARDT: Provide nitrogen and organic matter . . .

CULIK: Control erosion, by keeping the soil covered in fall and winter. Soil water-holding is improved when you plow down a cover crop.

HARWOOD: They've got to get erosion control. A cover crop will pick up soil nutrients in fall, and hold them so they're not washed away. There's much greater nutrient-retention efficiency.

THE NEW FARM: **And that's still more cost-effective than purchasing nitrogen?**

Dr. Richard R. Harwood

LIEBHARDT: I've been told that using green-manures or cover crops breaks even when nitrogen costs somewhere between 16 and 17 cents a pound. And if you grow your own vetch or sweet clover seed—whatever you're using—it would be even cheaper.

CULIK: Provided you get the biomass production from the cover crop. Sometimes, like the past two years here at the research center, we've had cool, wet spring weather and haven't gotten the spring regrowth we'd have liked. The farther north you go, the harder it may be to get good biomass production before the soil needs to be plowed.

HARWOOD: So it depends a lot on the climate.

LIEBHARDT: Right. You get somebody in North Dakota, that's going to be tough. If somebody's in a very cold growing region, they get their wheat in the ground and that's about all the time they've got. Overseeding is a bit like double-cropping, and you don't always have the moisture or the length of growing season to do it. For example, Warren Sahs, an agronomist at the University of Nebraska, said that they did better in their experiment with corn after soybeans than they did with corn after—I think—sweet clover. The legume was depleting moisture from the soil. That was the limiting factor.

HARWOOD: We can probably assume that the price of nitrogen is going to go out of sight in the next several years.

LIEBHARDT: A recent article in *Chemical and Engineering News* predicts another natural gas crisis by 1986. If that happens, I would think we could be looking at 35- to 40-cent nitrogen. It's interesting. In winter '84, I was at a meeting of the Office of Technology Assessment in Washington, D.C. A guy there from the Tennessee Valley

Authority has none of that programmed in his scenario. TVA is looking at 2- to 3-percent increases in energy prices in the short-term.
HARWOOD: Farmers need to be looking at cover crops and testing them, at least on a small acreage, even if they don't want to use them on their whole farm right away. Because within just a very few years, the price of nitrogen is going to force them to.
CULIK: In the North-central states, farmers want to plow in fall. Don Barnes, a USDA researcher who's studying alfalfa breeding and improvement at the University of Minnesota, is working on annual alfalfas that farmers can plow at the end of October. But the soil's open all winter, so it's not a good practice from an erosion-control standpoint.

Cold-Region Cover Crops

THE NEW FARM: **Is it absolutely necessary to include a legume in my rotation if I have no livestock?**
LIEBHARDT: Without it, there is no nitrogen.
THE NEW FARM: **But what do I do if I'm a cash-grain farmer in the Dakotas, or somewhere else where I can't overseed?**
LIEBHARDT: You have two choices. You can work it into your rotation. In other words, you could plant wheat and a vetch or alfalfa or sweet clover—something like that—and let it go a little longer in spring. Which means you will need to plant a shorter-season cash crop. Or you could purchase nitrogen.
CULIK: Is there any difference in the source of nitrogen? What about sewage sludge?
HARWOOD: The nitrogen content of sludge is pretty low. You're going to have to buy it, and nitrogen is expensive.
THE NEW FARM: **Is there a non-legume I could work into the rotation up there?**
LIEBHARDT: Oh sure, you could put in non-legumes.
HARWOOD: And probably mixtures, like Dick Thompson's using. He's in Iowa. He plants an oats-rye mix in September, and the oats winterkill while the rye regrows the next year. That keeps his ground covered, and gives him lots of biomass to plow down for his cash crop. He says the rye is allelopathic, too. It chemically inhibits the growth of some weeds. Thompson ridge-plants, though, so he has to be careful not to get too heavy regrowth or his planter sweep will get tangled. He sows about a bushel of oats and 10 pounds of rye per acre. The oats are dead by the next spring, so all he has to worry about is the rye.
LIEBHARDT: I think a mixture of rye and vetch, or rye and sweet clover, would work fine as far north as the Dakotas. After harvesting winter wheat in August, they could seed down the mixture. It would become established in fall. But it may take a considerable amount of time to produce much biomass—that's where your nitrogen is—in spring. It could also use to too much moisture, in which case it would actually reduce yields in the next crop. So they need to be careful and experiment a bit. I was talking with a farmer from North Dakota, and he asked me about overseeding. He said there's a big problem as soon as we discuss rotations, because his growing season is so short already.

CULIK: Well, there were fellows from North Dakota at a meeting I spoke at last year, and they were using cover crops. I think they were growing sweet clover in fallow years. They'd seed it into wheat, take off the wheat in late August or early September, then plow down the clover in fall or spring. If I remember correctly, some of them were overseeding annual ryegrass into sunflowers, too—either at last cultivation or at harvest.
HARWOOD: A good contact is the Earthcare group in Saskatchewan (The Earthcare Group, Box 1048, Wynyard, Saskatchewan, Canada, S0A 4T0). They wrote "Ecological Agriculture in Saskatchewan" (see "Earthcare Wrote The Book On Organic Farming," *The New Farm*, May/June '82). There's a loosely-knit farmers' organization up there. Some of them are cash-grain farmers, and they're pretty far north in Saskatchewan, actually. The Extension people at Regina, Saskatchewan, are researching a whole series of legumes, too.

Make The Most of Manure

THE NEW FARM: **Let's say I do have manure. How can I make the best use of it?**
CULIK: Storage. Apply it at the proper time to the proper crop. A roofed stable is needed to reduce the amount of water flowing into the manure pack, so nitrogen loss through leaching and runoff is minimized.
HARWOOD: A lot of people try to get a good carbon source into it. Dick and Sharon Thompson (*The New Farm* contributing editors) put all of their cornstalks in (see "Iowa State Discovers Nature's Ag School," *The New Farm*, Sept./Oct. '82). Ben Brubaker (who farms 320 acres adjacent to Rodale Research Center) puts straw in.
LIEBHARDT: To soak up the liquid. The urine is very important.
HARWOOD: Straw soaks up the nitrogen in urine, dries up the nitrogen so you don't lose it.
CULIK: In other words, have the manure packed until it gets thick enough that you have to dispose of it.
LIEBHARDT: I think the farmers who have gone to this liquid system have made their disposal problems worse. They have made a bulky product even bulkier.
HARWOOD: It saved them a little bit of work cleaning the barn, but, boy, it cost them. You add a tremendous amount of water. And when you do that, it becomes anaerobic and starts to stink. But farmers using liquid systems should still follow good manure-handling practices. They should apply and incorporate it as shortly before planting as possible. Or, they can spread it in fall, as long as they incorporate it right away.
THE NEW FARM: **First of all, we're assuming that they've got room to store manure.**
HARWOOD: Or that they've got their cropping system altered a bit so they can put it on at different times in the rotation. John (John Brubaker, Ben's son) has the problem in midsummer of where he's going to put it.
CULIK: He has to wait until after small grain harvest to apply it.
LIEBHARDT: Probably from about mid-May until the end of July, he's got no place to put his manure.
CULIK: That's why his barns are cleaned out, now.

LIEBHARDT: Then he'll go into the summer and put it on later.
HARWOOD: So he's probably putting it on some crops when he won't get optimum results.
LIEBHARDT: The spraying of manure on frozen soil isn't a good practice, either. You get runoff from it. If someone has a dairy farm, where they have to clean the barn every day, their main option is to run it out the end of the barn and stack it in a pile. Lots of straw, I think, is the answer to that.
HARWOOD: Otherwise, it's being exposed to denitrification, volatilization. You're going to lose 70 percent or so of your nitrogen. Well over half.

How Much Organic Matter?

THE NEW FARM: **A farmer from Michigan once told me his soil has too much organic matter. Five to 7 percent I think is what he said. He asks: Why should I add more organic matter to my soil when it's this high? Is it a problem having that much organic matter? How can I use it to my advantage?**

"It's hard to maintain organic matter content in the South, but that's not so crucial. You'll continue to get a nitrogen release if you're constantly adding new organic matter from green or animal manures."

LIEBHARDT: In soils where field crops are traditionally grown, I don't think you can have a problem with too much organic matter. I was talking to a North Dakota farmer, and he'd just put in new land that had never been plowed before. The organic matter content was 12.5 percent. And right across the field, where he'd been farming it for God knows how many years, it was 4.5 percent.
THE NEW FARM: **Then there's the South and the Southeast, where farmers have a hard time maintaining organic matter in the first place. What would you suggest for those guys, if they want to improve their soil without purchased inputs?**
LIEBHARDT: It's harder for them to maintain organic matter content, but that's not so crucial. Whether you have 1 percent or 2 percent or 5 percent, you'll continue to get a nitrogen release if you're constantly adding new organic matter from green or animal manures.

Overseeding: Why, Where And How?

HARWOOD: They get decent rainfall, so overseeding could really help them. Crimson clover does just beautifully down there. And vetch, as a winter crop.

LIEBHARDT: A lot of universities in the South have good research on overseeding: Auburn, Clemson's got some, Mississippi, Kentucky's got a lot.
HARWOOD: Anne Edwards, an Extension agent in King George County, Virginia, is a good contact.
CULIK: And the people at North Carolina State: Larry King and his graduate student.
HARWOOD: There's some work in Kentucky with hairy vetch.
LIEBHARDT: Wilbur Frye's got a lot of it.
HARWOOD: At University of Kentucky?
LIEBHARDT: Yep. It's written up in "Environmentally Sound Agriculture," William Lockeretz's book.
THE NEW FARM: **Suppose all a farmer has ever grown is corn and soybeans, and he'd rather not change. Can he still reduce his chemical costs?**
CULIK: Sure, certainly. Rotate them to reduce pesticide and fungicide costs. Use a legume cover crop, or overseed with a legume, to reduce purchased nitrogen.
LIEBHARDT: If he grew corn and soybeans every other year, he wouldn't need to use insecticides for corn rootworm.
THE NEW FARM: **No insecticide at all?**
LIEBHARDT: Absolutely not.
CULIK: Depending on where he is.
HARWOOD: That doesn't do a heck of a lot for weed problems, though. Some overseeding would help, there.
THE NEW FARM: **What would overseeding do for my weed problem? What crop would I overseed?**
LIEBHARDT: Depending on where you are, you could use vetch, sweet clover, crimson clover, rye . . .
THE NEW FARM: **Into corn and beans?**
LIEBHARDT: Yes. Or a mixture of grass and legumes—ryegrass.
THE NEW FARM: **Obviously, you'd have the same moisture constraints and growing season constraints we talked about earlier.**
HARWOOD: Again, it's almost like double-cropping. If you're in this intermediate zone—like from southeastern Pennsylvania through Ohio, Indiana, Illinois and Iowa—there's a whole band across there where you can't double-crop. It's real risky. But you *can* get a significant overseeded crop to grow. Of course, overseeding into soybeans still isn't refined. And then when you get down into the double-crop area, you're getting into winter crops. Some farmers plant soybeans after harvesting wheat, for example.
CULIK: There's a need to make the distinction between moisture limitations on establishing the overseeded species, and moisture depletion from the spring regrowth. In Nebraska and other Plains states, moisture depletion from regrowth can be a serious problem, particularly if it's a sod that has been in there for more than a year.
LIEBHARDT: That's right. In the western states, the depletion you're talking about is a serious problem. You can produce all kinds of nitrogen, but you suck up all the moisture and get nothing for your work.
THE NEW FARM: **Corn is irrigated there anyway, isn't it?**
LIEBHARDT: Well, a lot of it is, but most of it is still probably dryland.
HARWOOD: There's an old cropping systems map back in the '20s or '30s from the USDA that I've seen around, somewhere. It's from the "Yearbook of Agriculture," and

Dr. William C. Liebhardt (right) discusses research results with Martin Culik, who coordinated the Rodale Research Center's "Conversion Project."

it indicates cropping zones where particular kinds of crops fit. It shows where the dryland production areas are, where moisture will become limiting. And then if you're talking about spots where you can't quite double-crop, that's where overseeding fits.

LIEBHARDT: If you took the band from northern Delaware all through central Ohio, central Indiana, Illinois, Iowa—in that area and anything south of that line—overseeding should be no problem. It's north of there, generally, that it's risky, because biomass production will be more limited. Or, to get more biomass, you end up planting the main crop late.

THE NEW FARM: **What crops should I be thinking about if I want to overseed north of that line?**

LIEBHARDT: Well, if you're still growing corn and soybeans, you'd probably go with hairy vetch or sweet clover, something that makes it through the winter and gives you quick regrowth. You're going to be on the fringe, as far as the amount of biomass you'll get. There, a farmer's got to make a decision. Either he buys his nitrogen, or he lets his overseeded crop grow a little longer in spring, and maybe plants a shorter-season corn variety.

CULIK: Or let the cover crop grow as long as you can, then supplement. Use a smaller amount of nitrogen fertilizer than normal.

"It's very unlikely that you can't control weeds mechanically between the rows. I've rarely seen that."

LIEBHARDT: But if you let it go as long as you can, probably you're still going to be a little late in planting.

CULIK: But you'll still get some of the benefits from having that cover crop. I think the farmer who's growing corn and soybeans in a cash-grain system isn't likely to delay his planting.

HARWOOD: Again, talk with those Saskatchewan people, because they're right up there. They've got not only cold, but dry. And they're doing very well.

Get Wise To Weeds

THE NEW FARM: **Back to weeds. I can't quit using herbicides. If early rains keep me from cultivating, the weeds will take over.**

HARWOOD: That's right, until you get a rotation established.

LIEBHARDT: There you go again with the reduced use. Band over the row, and cultivate when you can. It's very unlikely that you can't control weeds mechanically between the rows. I've rarely seen that.

HARWOOD: Because you've got a long time frame in there. If it's wet, you've still got two or three weeks to control weeds between the rows. But you only have a few days to get the ones in the row. You have to hit it right.

LIEBHARDT: If they can't control weeds between the row with a cultivator, they've probably lost the crop to drowning. It would almost have to be that wet. I've just never seen a case where you can't get in and cultivate. And even if you can't get in early, you can always get the ones out in between.

HARWOOD: Dick Thompson argues that point all the time. He doesn't worry that much about weeds, as long as he can get control on the ridges. If you want to see an

example of how rotations help control different weeds, take a look at Ben's (Ben Brubaker) field right up here behind the research center. You can look out across 15 acres of wheat, and there isn't one single mustard weed on that hill. Yet if you look down from there a bit, there's an old alfalfa field that's been in a cool-season crop going on about the third year or so, and it's full of mustard. The difference is in the time of the rotation.
LIEBHARDT: Weeds are either cool-season or warm-season. A cool-season weed like mustard won't be a problem in crops like corn or soybeans, but might be in alfalfa. Likewise, a warm-season weed like pigweed can be a real problem in corn, but seldom in small grains or alfalfa. When you rotate, you keep changing the rules of the game.

Improve P and K Naturally

THE NEW FARM: **Let's shift gears a bit, away from nitrogen and weeds to P and K. Suppose a farmer feels rock phosphate and greensand don't supply P and K quickly enough. How can he be sure this year's crop will get enough of these nutrients without applying acidulated fertilizers?**
LIEBHARDT: First, he needs to get his soil tested to see if he even needs any P or K.
HARWOOD: He needs to have his soil tested and do some plant tissue tests to really get a true assessment. And then, secondly, he's got to minimize erosion loss no matter what he puts on. Soil erosion is going to cost him money in lost plant nutrients. One thing about phosphorus: It depends on your pH. If your pH is decent, don't lime the tar out of it so you get it up so high that rock phosphate won't do any good. If you're low on phosphorus and your pH is down, you can go a long way with rock phosphate. High pH— somewhere in the high 6s or 7s—is the only place where rock won't do any good. We've got our own farm out here where the native pH normally would have been in the high 5s or low 6s. We overlimed it and got it up to between 7 and 7.3. We put rock phosphate on and you can't even find it now.

"A cool-season weed like mustard won't be a problem in corn or soybeans, but might be in alfalfa. Likewise, a warm-season weed like pigweed can be a real problem in corn, but seldom in small grains or alfalfa. When you rotate, you keep changing the rules of the game."

THE NEW FARM: **What's an adequate pH to release the phosphorus in rock phosphate?**
LIEBHARDT: About 5.5 is best, if you have to pick one.
But most people don't want to farm at 5.5, so 6 to 6.5 is about right. You get over 6.5 or into 7, and forget it.
HARWOOD: And if you are up that high, you might be able to use rock in a compost pile. You get composting going, and run the rock through. When you get to the far West, those are recent volcanic soils and they have decent phosphorus anyway, in the loess soils.
LIEBHARDT: If you get West, let's say from Nebraska, the Dakotas or whatever, in the dryland areas, potassium is of almost academic significance. Rarely does anybody fertilize with potassium in those areas. On the other hand, they've got high pH, so the use of rock phosphate is a problem.
THE NEW FARM: **What do they do for P?**
LIEBHARDT: The only thing they can do is what most of them are doing: They're putting on triple. Of course, if they have animal manure, that's different.
CULIK: What could a cover crop provide as far as phosphorus?
LIEBHARDT: Dick Auld, a plant breeder at the University of Idaho, thinks Austrian winter peas have a real benefit with respect to phosphorus. If you look at Lambert's farm in eastern Washington (see "He *Nets* $60,000 A Year— Without Buying Fertilizer," *The New Farm,* March/April '83), where he cover crops with Austrian winter peas, he's got higher soil test values without phosphate fertilizer than the guys who use it.
THE NEW FARM: **Do you need that low pH to release phosphorus if you plow down a legume?**
LIEBHARDT: No, that wouldn't be the same principle. It's a different process, because with a legume, you're getting organic phosphorus instead of calcium phosphorus. Let's get back to potassium. That's something I haven't been comfortable with for awhile. Really, potassium 0-0-60 (muriate of potash) is a natural product. I know they don't like the chlorine, though.
HARWOOD: A good reference on potassium is the article Gene Logsdon wrote a few years ago (see "Potash: Weak Link In The Organic System," *The New Farm,* July/August '79). Gene went 'round and 'round on that one.
LIEBHARDT: Some organic farmers are using potassium sulfate, or they're using K-Mag, which is potassium magnesium sulfate. I think if we're going to help farmers use less acid fertilizers, we don't want to slam the door in their faces with this greensand. It's not a solution.

•

Hay As A Cash Crop

THE NEW FARM: **How can I think about taking land out of production when I'm having trouble making ends meet as it is? Should I be thinking of hay as a cash crop?**
CULIK: If there's a market. Be sure there's a market before you start growing it.
HARWOOD: Very few organic farmers take land out of production. I don't know of any who would do it. With overseeding, you don't have to.
CULIK: You don't always have to take land out of production with a cover crop, either. But if you're going to try and incorporate hay into a cash-grain system, then you have to have a market; you have to have labor for the machinery, and storage.

> *"If you can convince a banker or landlord to let you rotate, you can eliminate the insecticides for rootworm. Then, keep the herbicides if he insists on it. Or cut the rate down and band it, and show him that you can band plus cultivate cheaper."*

LIEBHARDT: Another option would be to grow a hay crop, and rent it out to a farmer who had animals for pasture. You can't always make those arrangements, but at least that's something you could consider. Or you might want to go out and buy some steers of your own, and just pasture them off and get rid of them. And when the pasture runs out, that's it.

Renters Can Reduce Costs, Too

CULIK: Let's deal with the situation where the farmer is leasing land, and he has a yearly or short-term lease. What can he do to reduce input costs?
HARWOOD: Number one: Convert his home acreage. In a lot of cases where you talk with those guys, their landlord or banker will be skeptical that it can be done.
CULIK: Well, with a one-year lease, they usually don't want to put in a hay crop or cover crop.
HARWOOD: They almost always have some home acreage. You convert that first, and make sure you have good, solid economic records on it so you can prove your point. More commonly, though, guys will have a longer-term arrangement, but the landlord won't let him reduce his chemical use 'cause he thinks it will lose him money. They'll be on a share basis or something like that.
CULIK: But if you're just going to incorporate the rotation and band herbicides, then there should be no problem.
HARWOOD: Yeah, but many people still won't go for it. We talked with a banker from Illinois, one time. Boy, that guy . . . they go right by the formula. Either it's corn or soybeans, and you do it this way. It had been several years since he'd even visited the farmers he serves, yet he was dictating how those guys farm. That's when economic proof comes in handy. We saw one farmer with a couple hundred acres. He was converting 40 of them, and what he needs to do is keep good economic records on it. Here's also where partial conversion would work. If you can convince the banker or landlord to let you rotate, you can eliminate the insecticide for rootworm. Then keep the herbicide if he insists on it. Or get him to let you cut the rate down and band it, and show him that you can band it—plus cultivate—cheaper. Not using the word "organic" helps, too, especially with bankers. Instead, tell them you're going to "structure the system to reduce inputs." Then go after them.
CULIK: The other thing I think we need to get across is that these farmers are going to have to work with the Extension service, particularly if they want all the cover crops that are suitable for their area—this type of information. We have to put more emphasis on the Extension service. Get them to service their clients.

Regenerating Chemically Treated Soil

THE NEW FARM: **What can I do on ground that's been treated with chemicals year after year?**
LIEBHARDT: If you're concerned about atrazine carryover, you could dig up a little bit of the soil and put some oats in it. You can do it in your kitchen window. If the oats survive, anything will survive.
CULIK: I think what he means is the biological activity in the soil. If you apply herbicides year after year, it may not be there.
THE NEW FARM: **What about a field that's been heavily fertilized with a synthetic product for 10 or 20 years? Can I anticipate any problems, any toxic buildup?**
HARWOOD: It's just going to take awhile until earthworm and biological activity resumes. Maybe an inoculum in the manure would help speed the process up. Growing a legume really helps, too. Biological activity really jumps when you grow alfalfa or clover. The point here is that farmers need to be looking not only at today's economics, but at what they're going to be five or 10 years down the road. And, you know, it's really clear: Farmers better darn well start gearing up for some of these practices, because if—when—the price of nitrogen goes up, they're going to be locked into a system and won't know how to cut their costs.
CULIK: I would say, along with that, that management has to be increased. When you're not depending on fertilizers or pesticides, you have to be much more of a student of your fields.
HARWOOD: So the farmer needs to spend more time planning.
CULIK: It's not a recipe anymore. This is a biological system, and the farmer is part of it.

Equipment: Plan Before You Buy

HARWOOD: One thing we haven't talked about is equipment. In the *New Farm* survey, we asked some questions about what piece of equipment you buy first. A lot of farmers are arguing the chisel plow, and the rotary hoe, to get the weeds down in the rows. You can't have too many false starts in your cropping system, though, because you can't keep going out and buying machinery. So you really

> *"Farmers need to be looking not only at today's economics, but at what they're going to be five or 10 years down the road."*

Aerial view of 72 test plots in Rodale Research Center's five-year Conversion Project.

> "You don't have to spend a fortune just to try some of these things. You can rent or borrow a piece of equipment."

need to get out and talk with some people, especially other farmers.

THE NEW FARM: **If I'm a corn and soybean grower, and you're telling me to overseed, what's that going to do to my equipment needs?**
HARWOOD: You can hire it out, if you just want to try it. Or you can buy a hand rig and do it on five acres before you gear up. You don't have to spend a fortune just to try most of these things. Or even to rent or borrow a piece of equipment.
CULIK: The instance where Diane's (RRC Entomologist Diane Matthews) landlord went out and bought a cultivator: Now you feel pretty good about it, because almost certainly it's going to work.
LIEBHARDT: Yeah, if he sets the thing up right.
HARWOOD: Especially if it's rear-mounted. More and more people go with the rear-mounted these days. It used to be front-mounted, because there was more precision. But people are rotary-hoeing, and they don't demand the precision anymore that we used to need.

THE NEW FARM: **Should we be considering ridge tillage just as a really important option that the farmers can consider?**
HARWOOD: I think so. Certainly a coming option in the wet soil, where you have to fall plow. That's the big advantage. If you have light, sandy soil, then there are two reasons you wouldn't do it: First, you could work the soil wet in the spring; and second, the ridges wouldn't stay up anyway. With a lot of these things that we've been talking about, I don't see that there's much size and scale limitation, except for management. You start running into management problems where you're too big to be able to manage it. You know another point that needs to be made somewhere in here are the economics. If you look at Eddie Steiger (a cash-grain farmer near the RRC), his up-front cash costs on that corn planted out there are probably somewhere between $100 and $150 an acre, because he's got monocrop corn. He's putting everything down at planting time: his herbicide, insecticide, his fertilizer, everything right at planting time. And he's got to put up something like $150 cash per acre for cost right off the top. Now with the Brubaker system, that cash input— the cash flow—is a lot different. It comes on at different times, so the credit requirements are quite a bit less, and a lot of it is labor. Even though Brubaker's total production cost may be 10 percent less, the credit cost is far less.
CULIK: He borrows primarily when he buys cattle.
THE NEW FARM: **But up-front production costs?**
CULIK: Seed.
HARWOOD: That's one of the changes that we're talking about in the system, just going to these lower inputs.

The Dollars And Sense Of Resource-Efficient Farming

The Rodale Research Center's Conversion Project. How we did it, what we learned, and what it means to you.

"I'd really like to reduce my costs, but I can't afford the reduction in yields—and income—for those first few years."

We heard that familiar lament so often at the Rodale Research Center (RRC) that we decided to tackle the problem head-on by studying what happens to a farm's yields, soil and income when chemical use is abruptly stopped. We were intrigued by studies like the USDA's 1980 "Report and Recommendations on Organic Farming," which shows that well-established organic farms have yields similar to neighboring farms, while new ones often suffer three to five years of reduced yields. Our experiment proved this to be true. On plots where we suddenly quit using chemicals, corn grain yields plunged by 40 percent the first year. By the fourth year, organic corn grain yields were still 8.5 percent below those on chemical plots—but the organic corn was earning more money.

Few farmers can wait that long for yields and income to rebound. If we were to be of any help to the growing number of North American farmers interested in using fewer chemicals, then we had to find a way for them to get through those critical first few years of reduced inputs without any economic sacrifice.

To do that, we designed three, five-year rotations: *Organic Farming With Animals, Organic Farming Without Animals* and *Conventional Cash Grains*. Although most farmers would probably reduce chemicals gradually, we cut them off altogether in the organic plots, to confront the problems at their worst.

Here's a quick summary of the important lessons we have learned so far:

Cut Costs, Not Income. Our farming system that simulates a diversified, organic livestock operation makes the most money, with an average annual return over variable costs of $174 per acre. That's $30 per acre more than the average for our conventional corn and soybean cash-grain rotations. Our organic cash-grain system—which uses legume plowdowns as the only source of nitrogen—returns somewhat less than conventional cash grain: $125 per acre.

Start With Soybeans, Small Grains And Legume Hays. When we grew corn without purchased N the first year, yields were up to 40 percent less than conventional corn. However, soybeans, which fix their own nitrogen, yielded essentially the same in the organic and conventional systems.

One strategy that is particularly well-suited for our area in southeastern Pennsylvania is to plant legume hay with a small grain nurse crop. The small grain shelters the legume for good stand establishment, and yields a profitable grain harvest the first year. Depending on weather, a cutting of hay is also possible the first year. The thick ground cover of the two crops helps smother weeds, reducing future weed pressure.

Small grains overseeded with legumes can help minimize nitrogen and weed problems, and maintain cash flow during the first few years that chemicals are being reduced.

Rotations Are The Key To Success. The biological diversity in an established rotation is essential for maintaining soil fertility, reducing weed competition, and controlling insects. Even in the conventional system, we noticed slightly higher yields in corn following soybeans than in corn following corn when the same tillage practices were used.

It Takes Time. The benefits of crop rotation don't show up right away. Establish rotations one field at a time. Learn from your initial successes and failures, and apply them to the next field.

Sharpen Your Management Skills. Become familiar with the management of crops that might not have been grown in your area since your childhood. Acquaint yourself with the subtleties of mechanical weed control. Good timing is essential. Know and thoroughly understand your soils and the nutrient needs of your crops. Field operations on a rotation-based farm are more spread out. You may be able to avoid the pressure that results when you have one week to get 400 acres of corn planted. Diversification takes lots of planning and a considerably different work schedule.

There Are No Cookbook Recipes. What worked for us may not be suited for your unique situation. Keep flexible; see what works and what doesn't, and respond accordingly as the biological life of your crops and soil changes. *Adapt* what you can from our research and the experience of others; select crops and management techniques that suit your rotation, climate, soils, markets, labor, equipment and banker.

At First, Some Chemicals May Be Needed. Use them if you need to—don't if you don't. Judicious use of chemicals can help smooth over some economic stumbling blocks. For example, some supplemental nitrogen may be needed if you didn't get the big legume plowdown you'd hoped for. Soils not naturally high in P and K may need additions of these nutrients. Spot spraying herbicides where weeds have been particularly troublesome will help clear the way for better mechanical weed control in the future.

Designing The Conversion Experiment

Our first job was to find some land that had been farmed chemically. This was a problem, since the 305 acres at the RRC has been farmed organically for more than a decade. But our neighbors to the west farm 2,000 acres of corn, soybeans and wheat, using standard insecticides, herbicides and fertilizers. We rented their adjoining 15-acre field where corn and wheat had been grown conventionally for several decades. In 1979, the field was in corn for grain, followed by winter wheat. Wheat was harvested in summer 1980, and the field was idle until we began our project in March 1981. During this time, giant foxtail (*Setaria faberi*) took over and set seed ("A 20-bushel-per-acre crop," jokes one of our agronomists).

We decided to test three different five-year rotations. The first, *Organic Farming with Animals*, simulates an integrated, organic operation where field crops are raised mainly for beef or dairy cattle. Limited amounts of manure are available. Corn silage and hay are grown. System number 2, *Organic Farming Without Animals*, simulates a cash-grain farm without access to manure. The only source of nitrogen is legume plowdowns. A marketable grain crop is harvested each year. The third system, *Conventional Cash Grains*, uses chemical fertilizers and herbicides at rates recommended by Pennsylvania State University. (We haven't needed any insecticides yet. The corn-soybean rotation has discouraged any infestations that might reduce yields.) This is the control, and provides a comparison with our organic rotations.

We suspected that some crops, like soybeans and legume hays, might be better to start with than corn, so we began each of the three rotations at three different points, giving us a total of nine treatments. In scientific jargon, a treatment is an experimental condition in which only one practice, like fertilizer rates, is varied. To keep track of all

Conversion Project Rotations

Organic Farming With Animals (System 1)

Treatment	1981	1982	1983	1984	1985
1-1	Spring Oats / Red Clover Hay	Red Clover Hay	Manure / Corn	Soybeans	Manure / Corn Silage
1-2	Corn	Soybeans	Manure / Corn Silage	Wheat / Red Clover OS*	Red Clover Hay*
1-3	Manure / Corn Silage	Wheat / Red Clover OS*	Red Clover Hay	Manure / Corn	Soybeans

Organic Farming Without Animals (System 2)

Treatment	1981	1982	1983	1984	1985
2-1	Spring Oats / Red Clover Hay	Corn	Spring Oats / Red Clover Hay	Corn	Soybeans
2-2	Soybeans	Spring Oats / Red Clover Hay	Corn	Wheat / Hairy Vetch OS*	Corn
2-3	Corn	Soybeans	Spring Oats / Red Clover Hay	Corn	Spring Oats / Annual Legume OS*

Conventional Cash Grains (System 3—Control)

Treatment	1981	1982	1983	1984	1985
3-1	Corn	Corn	Soybeans	Corn	Soybeans
3-2	Soybeans	Corn	Corn	Soybeans	Corn
3-3	Corn	Soybeans	Corn	Corn	Soybeans

*Overseeding

the treatments, we gave each a two-number code. The first number indicates the farming system, the second tells the entry point into the rotation. For example, treatment 2-1 is farming system 2 (*Organic Farming Without Animals*), entry 1 (the rotation starting with spring oats followed by red clover).

This coding is necessary because each treatment is repeated eight times. We divided the field into 72 plots, each 60 feet wide and 300 feet long. These large plots allow us to use full-scale field equipment, just as a farmer would. We left 5-foot grass buffer strips between treatments to reduce the movement of fertilizers and pesticides, and 35-foot grass roadways at the ends of the plots for field equipment traffic. Only the center rows in each plot were harvested, to avoid the border effect. We analyzed soil samples to check for uniformity in the field, and to give us good baseline data to monitor changes in soil fertility.

In addition to heavy weed pressure, we also went into our first season with lower-than-normal soil moisture. There was a severe drought during the summer and fall of 1980, and below-normal recharge during the winter of 1980-81. On March 26, 1981, the entire field was moldboard plowed and we were ready to start.

We recorded the cost of seed, fertilizer and herbicides for each treatment, and determined field operation costs using the booklet "Pennsylvania Machinery Custom Rates," published yearly by the Pennsylvania Department of Agriculture. (If anything, that would favor the conventional treatments, because they had fewer operations.) All grain and hay crops were "sold" at local prices, as reported at harvest in *The Lancaster Farmer* and the *Pennsylvania Agriculture Digest*.

The Bottom Line

Despite severe droughts in 1981 and '83, treatment 1-3 had the best four-year average return over variable costs: $218 per acre. Part of this success must be attributed to the seller's market for silage in 1981. The second best treatment, 1-1, returned an average of $183 per acre. It was least affected by the 1981 drought.

Conventional Cash Grain rotations suffered most in the 1981 drought. Soil crusting caused spotty corn germination and highly variable populations averaging around 13,000 plants per acre. If left till harvest, differences in corn grain yields would have resulted from different plant populations, not different treatments. A farmer might have left the spotty stand. We needed to gather some solid scientific data, so we replanted all corn plots.

That delay increased the affects of the drought. The corn in treatment 1-3 produced a lot of biomass and good silage yields (11.96 tons per acre). But the drought in July, August and September severely reduced grain yields in all of the other corn plots. Yields ranged from 22 to 40.6 bushels per acre. Data from "The Kutztown Farm Report" on the neighboring Brubaker farm, and state and county averages for 1981, indicate that if the corn had been planted on time, our yields would have been similar to those in 1983 (78.1 to 101.6 bushels per acre), a year with almost identical moisture conditions. Had this been the case, treatments 3-1 and 3-3 would have had average four-year returns in the same range as our best organic treatment.

The drought and replanting similarly affected the economic returns for treatment 1-2. It is doubtful that yields and returns from this treatment would have been compara-

When manure was available, two of the organic rotations had the highest average returns per acre. Starting the rotation with a small grain/legume hay helped boost long-term earnings without manure.

Organic Farming With Animals

			Actual cash returns per acre		
Treatment	1981	1982	1983	1984	4-yr. avg.
1-1	Oats/Red Clover $128	Red Clover $121	Corn Grain $249	Soybeans $235	$183
1-2	Corn Grain −$27	Soybeans $154	Corn Silage $256	Wheat-Red Clover $102	$122
1-3	Corn Silage $243	Wheat-Red Clover $196	Red Clover $132	Corn $301	$218

Organic Farming Without Animals

2-1	Oats/Red Clover $139	Corn $111	Oats/Red Clover $115	Corn $247	$153
2-2	Soybeans $83	Oats/Red Clover $49	Corn $232	Wheat-Red Clover $96	$115
2-3	Corn −$49	Soybeans $155	Oats/Red Clover $137	Corn $179	$106

Conventional Cash Grain

3-1	Corn −$89	Corn $187	Soybeans $195	Corn $280	$144
3-2	Soybeans $77	Corn $163	Corn $188	Soybeans $208	$159
3-3	Corn −$82	Soybeans $120	Corn $215	Corn $262	$128

ble to the 1983 corn grain yields in treatment 1-1, however, since 1-1 followed two years of red clover and a manure application. Instead, yields would probably have been in the 60- to 70-bushels-per-acre range.

The *Organic Farming Without Animals* system did not perform as well economically as the other organic system. Based on the four-year results, a rotation beginning with a small grain with legume overseeding seems to be the best way to overcome nitrogen deficiencies when manure is not available.

Few, if any, changes were noticed in the soil during the first four years of the Conversion Project.

Phosphorus levels were from eight to 12 times higher than necessary for corn in the top 8 inches, and remained high throughout the experiment. Potassium levels were twice what is sufficient for corn, and increased in treatments that received manure. K levels in the conventional system, which received from 10 to 30 pounds of potash fertilizer yearly, were no higher than in the organic system.

Soil conditions at the start of the Conversion Project. Samples taken in spring 1981; from 0 to 6 inches deep.

Soil pH	6.64	Magnesium	321.6 lb/A
Buffer pH	6.80	Calcium	2984 lb/A
Nitrate Nitrogen	24 lb/A	CEC	11.72
Phosphorus	172.7 lb/A	Organic Matter	2.44%
Potassium	429 lb/A		
Soil types	88% Comly Silt Loam (a Typic Fragiudalf, fine-loamy, mixed mesic)		
	12% Berks Shaly Silt Loam (a Typic Dystrochrept, loamy-skeletal, mixed mesic)		
	Duffield Silt Loam (an Ultic Hapludalfs, fine-loamy, mixed mesic)		

Farm Manager Tom Morris (foreground) and Agronomist Martin Culik prepare deep core samples for analysis. P and K remained sufficient in organic and conventional plots throughout the experiment.

Conversion Project Rotations

Organic Farming With Animals (System 1)

Treatment	1981	1982	1983	1984	1985
1-1	Spring Oats Red Clover Hay	Red Clover Hay	Manure Corn	Soybeans	Manure Corn Silage
1-2	Corn	Soybeans	Manure Corn Silage	Wheat Red Clover OS*	Red Clover Hay
1-3	Manure Corn Silage	Wheat Red Clover OS*	Red Clover Hay	Manure Corn	Soybeans

Organic Farming Without Animals (System 2)

	1981	1982	1983	1984	1985
2-1	Spring Oats Red Clover Hay	Corn	Spring Oats Red Clover Hay	Corn	Soybeans
2-2	Soybeans	Spring Oats Red Clover Hay	Corn	Wheat Hairy Vetch OS*	Corn
2-3	Corn	Soybeans	Spring Oats Red Clover Hay	Corn	Spring Oats Annual Legume OS*

Conventional Cash Grains (System 3—Control)

	1981	1982	1983	1984	1985
3-1	Corn	Corn	Soybeans	Corn	Soybeans
3-2	Soybeans	Corn	Corn	Soybeans	Corn
3-3	Corn	Soybeans	Corn	Corn	Soybeans

*Overseeding

A THREE-YEAR LOOK AT WEATHER, FIELD OPERATIONS, CROP RESPONSES AND ECONOMICS

1981

Weather—Temperatures during the spring were slightly above normal, while in August, September and October, they were slightly below normal. Precipitation during July, August and September was about 5 inches below normal.

Field Operations—The beef cattle manure for treatment 1-3 (corn silage) was applied at a rate of 4 tons per acre (dry weight) on May 4, and disked within 24 hours. Starter fertilizer (10-30-10) was applied to the conventional corn. Atrazine (1.2 quarts per acre) and alachlor (Lasso, 1.75 quarts per acre) were applied to the conventional corn on May 14.

Soil in all corn plots crusted over during the two weeks following planting. On May 22, all corn was rotary hoed. Emergence was spotty, so corn was replanted on June 7. Weed control in the organic corn consisted of one rotary hoeing and three row cultivations.

A dry powder inoculum was applied to all soybeans prior to planting. Alachlor (Lasso) and metribuzin (Sencor) were applied to the conventional soybeans (treatment 3-2) on May 30 at a rate of 2 quarts per acre and 0.75 pounds per acre, respectively. Organic soybeans (treatment 2-2) were row-cultivated twice.

Crop Responses—We encountered many problems with corn in both the organic and conventional plots. Replanting due to poor initial stands, lack of soil moisture, and a Northern corn rootworm (*Diabrotica longicornis*) infestation combined to reduce yields to 30 percent of the potential for our soils. Even then, conventional corn yielded significantly better.

With such results, we could see why some farmers give up on organic methods after the first year. Our poor showing was due in part to our lack of familiarity with the field. We overworked it before planting. That, coupled with the peculiar spring weather, led to the crusting. Had this not happened, we could have used a longer-season

Organic corn silage provided the best cash returns in 1981.

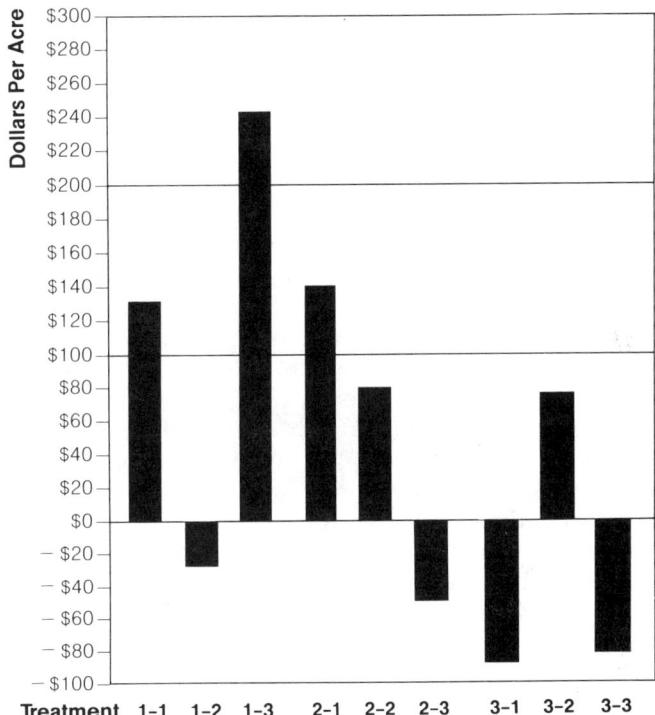

1981 Field Operations/Drought hurt all yields. Despite that, the two organic systems had the best cash returns.

Treatment	Crop	Planting Date	Rate[1]	Weed Control	N-P$_2$O$_5$-K$_2$O (lb/A)	Yield (bu/A)	Returns Over Variable Costs
				Cultivations	Manure		
1-1	Oats/Red Clover	3/26	48/10	-	-	60/1.2 tons	$128
1-2	Corn (105-day)	6/7	25,400	4	-	25	-$27
1-3	Corn Silage	6/7	25,400	4	160-51-206	12.0 tons	$243
					Plowdown		
2-1	Oats/Red Clover	3/26	48/10	-	-	66/1.2 tons	$139
2-2	Soybeans	5/26	177,000	2	-	26	$83
2-3	Corn (105-day)	6/7	25,400	4	-	22	-$49
				Herbicides	Fertilizer		
3-1	Corn (105-day)	6/7	25,400	atrazine alachlor dicamba	162-29-10	38	-$89
3-2	Soybeans	5/26	177,000	alachlor metribuzin	-	26	$77
3-3	Corn (105-day)	6/7	25,400	atrazine alachlor dicamba	162-29-10	41	-$82

[1]Rate is pounds per acre for cereals and red clover, and seeds per acre for corn and soybeans.

Better weed control in both the conventional and organic systems would have improved 1981 corn yields.

Treatment	Grain Yield (bu/A)	Leaf Tissue Nitrogen (%)	Broadleaf Weed Biomass (lb/A)	Grass Weed Biomass (lb/A)
1-2	24.9	1.98	196.7	592.7
1-3	Silage[1]	2.41	179.5	566.9
2-3	22.0	2.03	194.0	544.7
3-1	38.0	2.87	55.2	283.0
3-3	40.6	2.95	24.9	235.0

[1] Silage yields averaged 11.96 tons per acre.

hybrid. We wouldn't have hit the drought at the critical time of grain formation, or the rootworm at the peak of its second generation. The result would have been more respectable yields.

Still, we learned a lot from our mistakes. For example, leaf tissue tests showed nitrogen deficiencies in the organic corn. The corn silage (treatment 1-3) was less deficient than the other two organic treatments (1-2 and 2-3), probably due to the manure it received before planting. The conventional corn did not appear nitrogen-deficient.

The differences in leaf tissue nitrogen between the conventional corn and the organic corn silage seem puzzling, since each received the same analysis of fertilizer, although in different forms. In this case, the manure didn't perform as well as the chemical. We speculate that there are several reasons for this. Manure was applied early in the season, and the readily available nitrogen may have been leached away in the wet June weather. Then, during the dry season, there was insufficient moisture for the microbial action necessary to release more nitrogen from the manure. In addition, a large amount of highly carbonaceous foxtail residue may have tied up the manure nitrogen that was released. These residues may have actually helped some of the conventional corn plots by tying up some of the chemical nitrogen and preventing it from leaching away.

Weed pressure was significantly higher in the organic corn, but was also heavy in the conventional plot, probably due to replanting corn after the original herbicide application. Yields correlated better with leaf tissue nitrogen than with weed biomass, which means nitrogen—not weeds—was limiting the organic corn grain yields. However, some of the nitrogen deficiency could have been caused by weed competition. Better weed control in both conventional and organic plots may have produced higher yields. Had the season been wetter, favoring weed growth and making more nitrogen available, weeds may have been more of a problem.

Soybeans were a different story. They were grown in one organic and one conventional treatment. No fertilizer nutrients were applied to either treatment, and the only management difference was the use of herbicides on the conventional soybeans. Nodulation was poor, apparently the result of a low rhizobia population in the soil (soybeans had not been grown in this field for many years), and inadequate inoculation with a dry powder inoculum. Good nodulation would not have favored either treatment.

Conversion Project Rotations

Organic Farming With Animals (System 1)

Treatment	1981	1982	1983	1984	1985
1-1	Spring Oats / Red Clover Hay	Red Clover Hay	Manure / Corn	Soybeans	Manure / Corn Silage
1-2	Corn	Soybeans	Manure / Corn Silage	Wheat / Red Clover OS*	Red Clover Hay
1-3	Manure / Corn Silage	Wheat / Red Clover OS*	Red Clover Hay	Manure / Corn	Soybeans

Organic Farming Without Animals (System 2)

Treatment	1981	1982	1983	1984	1985
2-1	Spring Oats / Red Clover Hay	Corn	Spring Oats / Red Clover Hay	Corn	Soybeans
2-2	Soybeans	Spring Oats / Red Clover Hay	Corn	Wheat / Hairy Vetch OS*	Corn
2-3	Corn	Soybeans	Spring Oats / Red Clover Hay	Corn	Spring Oats / Annual Legume OS*

Conventional Cash Grains (System 3—Control)

Treatment	1981	1982	1983	1984	1985
3-1	Corn	Corn	Soybeans	Corn	Soybeans
3-2	Soybeans	Corn	Corn	Soybeans	Corn
3-3	Corn	Soybeans	Corn	Corn	Soybeans

*Overseeding

The organic soybean plots looked weedy. Our measurements showed nearly 10 times more grass weed biomass. However, yields were essentially the same as in the conventional system: 26 bushels per acre.

Better spring weather would have allowed us to cultivate more often for better weed control in the organic soybeans. While the weeds didn't reduce yields, they did set seed and increase weed pressure for future seasons. Had the summer continued to be wet, weed competition may have been more of a problem. Until the rotation effect has a chance to suppress weeds, cultivation is your only form of non-chemical weed control.

Two organic plots had spring oats sown with red clover. Management was identical; there were no weed problems or differences in grain or hay yields. Average oat grain yields in our area are about 50 bushels per acre; both organic plots beat that by 10 bushels. Hay yields of 1.16 and 1.15 tons per acre were limited by drought in July and August.

Economics—Poor weather that reduced crop yields also increased market prices, favoring the two organic systems.

Corn grain production lost money in all cases, but less in the organic systems where input costs were lower. Organic corn silage gave good returns due to low input costs and high market prices. Organic soybeans returned slightly more than conventional soybeans, which bore the added cost of herbicides.

1982

Weather—Except for a cool, wet June, weather was favorable. Temperatures were generally cooler than normal, while precipitation was above average, particularly in June and August.

Field Operations—Red clover was sown by hand into wheat plots to simulate a tractor-broadcasted seeding. Red clover in treatment 1-1 produced two hay cuttings. Weather delayed the first cutting till seven days past full bloom; the second cutting was rained on before baling. The oat stand (treatment 2-2) was reduced by late planting and competition from giant foxtail. Red clover in treatment 2-1 was plowed down on May 13 when it was about 12 inches high.

Corn stands were irregular due to poor emergence and pheasant damage, mainly in the organic plots. Late planting and no starter fertilizer may have also contributed to the lower populations in the organic corn.

Corn in treatment 2-1 received one rotary hoeing and two row cultivations. Conventional corn plots were tilled differently because of previous crop histories. Treatment 3-1 had corn in 1981 and was moldboard plowed, while treatment 3-2 had soybeans in 1981 and was chisel plowed. Chiseling is good for most fields in our area, but this field was so wet that we gave up chiseling after 1982. Both treatments were disked and cultipacked. Starter fertilizer (15-30-30) was band-applied to conventional corn at planting. This corn was later sidedressed with urea at 110 pounds of nitrogen per acre. Alachlor (Lasso) and atrazine were applied on May 14 at 0.9 and 1.0 quarts per acre, respectively.

Four of the soybean plots were planted late (June 26 as opposed to June 10 for the others) because of wet field conditions, and at a slightly lower population (160,000 seeds per acre instead of 175,000). In all soybean treatments, a granular, soil-applied inoculum was used to introduce rhizobia. Alachlor (Lasso) and dinoseb (Premerge) herbicides were applied at the rate of 2 quarts each per acre on June 18 to the early-planted conventional soybeans, and on July 2 to late-planted beans. Organic soybeans received three row cultivations.

1982 Field Operations/*One of the* Organic Farming With Animals *treatments again earned the most.*

Treatment	Crop	Planting Date	Rate[1]	Weed Control	N-P_2O_5-K_2O (lb/A)	Yield (bu/A)	Returns Over Variable Costs
				Cultivations	Manure		
1-1	Red Clover	2nd year		-	-	1.7 tons	$121
1-2	Soybeans	6/10	175,000	3	-	49	$154
1-3	Wheat-Red Clover	9/21/81 3/18/82	120 10	-	-	39/2 tons	$196
					Plowdown		
2-1	Corn (100-day)	5/14	23,000	3	121-25-145	88	$111
2-2	Oats/Red Clover	4/24	48/10	-	-	32/0.6 tons	$49
2-3	Soybeans	6/10	175,000	3	-	49	$155
				Herbicides	Fertilizer		
3-1	Corn (110-day)	5/11	23,000	alachlor atrazine	125-30-30	153	$187
3-2	Corn (110-day)	5/11	23,000	alachlor atrazine	125-30-30	137	$163
3-3	Soybeans	6/10	175,000	alachlor dinoseb	-	43	$120

[1] Rate is pounds per acre for cereals and red clover, and seeds per acre for corn and soybeans.

Crop Responses—Differences in plant populations affected corn grain yields to some extent, favoring treatment 3-1. Treatment 3-2 had a lower population due to the different tillage (chisel instead of moldboard) followed by cool, wet weather. The organic corn (treatment 2-1) population was lower due to cultivation and pheasant damage. With a higher population and a longer-season hybrid, we may have been able to stretch organic corn yields into the 120-bushel range.

In 1982, cash returns for organic wheat seeded to red clover were about the same as for conventional corn.

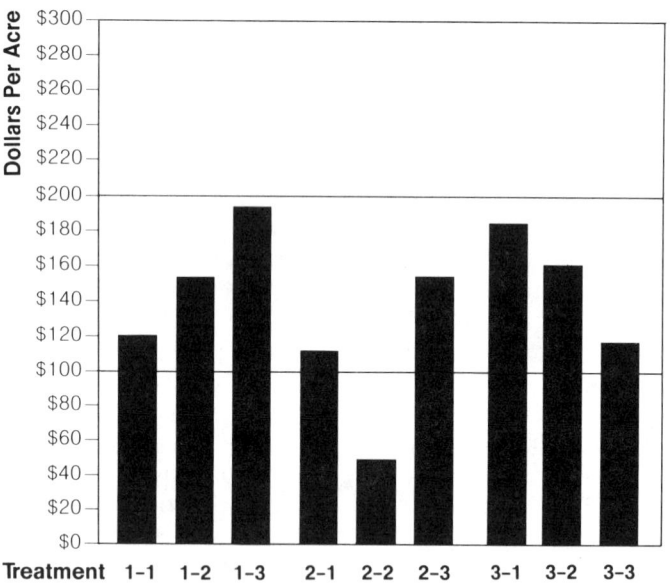

Leaf tissue analysis showed sufficient nitrogen for all corn. Although the organic corn following red clover plowdown (treatment 2-1) had slightly lower N levels than the conventional corn, it was not significantly different in the statistical analysis. These lower levels were probably the result of weed competition.

Severe weed competition was probably the most important reason for the lower organic yields. Giant foxtail was the dominant weed, mostly in the rows. Pigweed (*Amaranthus*), lambsquarters (*Chenopodium*) and bindweed (*Convolvulus sepium*) were also common.

While yield differences between the conventional treatments were not significant, plant growth differences—height, stand, and color—were observed throughout the season. Corn following soybeans usually does better than corn following corn, but the difference in tillage and the cool, wet weather early in the season favored the corn after corn.

All three soybean treatments followed corn. The lower plant population in the conventional soybeans (treatment 3-3) was due to injury from residual corn herbicide. The plants had crinkled leaves and stunted growth early in the season, which they eventually outgrew. While we were following Pennsylvania State University recommendations in the conventional system, it became clear that our shaly soils required lower herbicide rates. Rates were reduced in future seasons. Statistics showed a strong possibility that the lower populations in the conventional plots reduced yields. All soybeans had good nodulation with the granular, soil-applied inoculum.

Broadleaf weed biomass was higher in the organic treatments, but grass weed biomass was not significantly different. Weeds weren't a problem in the organic soybeans, because the weather allowed timely cultivations.

Oat grain yields were 32.3 bushels per acre. Once again, there was no weed competition, but yields were lower

Conversion Project Rotations

Organic Farming With Animals (System 1)

Treatment	1981	1982	1983	1984	1985
1-1	Spring Oats Red Clover Hay	Red Clover Hay	Manure Corn	Soybeans	Manure Corn Silage
1-2	Corn	Soybeans	Manure Corn Silage	Wheat Red Clover OS*	Red Clover Hay
1-3	Manure Corn Silage	Wheat Red Clover OS*	Red Clover Hay	Manure Corn	Soybeans

Organic Farming Without Animals (System 2)

2-1	Spring Oats Red Clover Hay	Corn	Spring Oats Red Clover Hay	Corn	Soybeans
2-2	Soybeans	Spring Oats Red Clover Hay	Corn	Wheat Hairy Vetch OS*	Corn
2-3	Corn	Soybeans	Spring Oats Red Clover Hay	Corn	Spring Oats Annual Legume OS*

Conventional Cash Grains (System 3—Control)

3-1	Corn	Corn	Soybeans	Corn	Soybeans
3-2	Soybeans	Corn	Corn	Soybeans	Corn
3-3	Corn	Soybeans	Corn	Corn	Soybeans

*Overseeding

Weeds, not nitrogen deficiencies, reduced organic corn yields in 1982.

Treatment	Population plants/acre	Grain Yield (bu/A)	Leaf Tissue Nitrogen (%)	Broadleaf Weed Biomass (lb/A)	Grass Weed Biomass (lb/A)
2-1	17,600	87.5	2.60	178	2058.6
3-1	20,670	152.9	2.74	13.4	61.4
3-2	18,940	137.4	2.66	—	—

than in 1981, due to poor weather and late planting. Red clover hay yields were also reduced to 0.6 tons per acre.

The second-year red clover in treatment 1-1 was cut twice for hay with a total yield of 1.7 tons per acre. Few weeds were observed until after the second cutting, when giant foxtail began to increase.

Wheat yields in treatment 1-3 were 39 bushels per acre, while red clover hay yielded 2 tons per acre. There were no plant nutrient deficiencies and few weeds.

Economics—Treatment 1-3 gave the best cash return of the season: $196 per acre.

Despite lower production costs for organic corn, conventional corn yields were enough higher to provide a better net return per acre.

Soybean production costs were similar, because we used less expensive herbicides in the conventional plots, and cultivated more in the organic. Organic soybeans made more money because yields were 29 percent higher.

1983

Weather—The season started off cool and wet in April, May and June. Then it turned hot and dry in July, August and September, severely stressing crops and reducing yields.

Field Operations—Beef cattle manure was applied to treatment 1-2 at a rate of 3.2 tons per acre (dry weight) on November 19, 1982, and plowed the same day. Red clover in treatment 1-1 was so sparse in spring that no hay cutting was taken. Beef cattle manure was spread on this treatment on May 6 at a rate of 3.4 tons per acre. Red clover in treatment 2-2 was plowed down on May 10 and yielded an average of 1.3 tons per acre of dry matter. One rotary hoeing and two row cultivations were used to control weeds in the organic corn.

Only one cutting of red clover hay in treatment 2-1 and 2-3 was possible because of the drought. In treatment 1-3, we cut red clover twice, but the second time it was starting to bloom at a height of only 13 inches.

A starter fertilizer (10-30-10) was banded on the conventional corn at planting. Cyanazine (Bladex) and metolachlor (Dual) herbicides were applied on May 12 at 1.4 and 0.7 quarts per acre, respectively. The conventional corn was sidedressed with ammonium nitrate at 110 pounds of nitrogen per acre.

Granular, soil-applied inoculum was used with the conventional soybeans. They were sprayed with metolachlor (Dual) and linuron (Lorox) herbicides at rates of 0.8 and 0.4 quarts per acre, respectively.

Crop Responses—Once again, cultivation and pheasant damage reduced organic corn populations. Yields in treatment 1-1 were only 10 percent less than the conventional

1983 Field Operations/For grain or silage, organic corn made the most money.

Treatment	Crop	Planting Date	Planting Rate[1]	Weed Control	N-P₂O₅-K₂O (lb/A)	Yield (bu/A)	Returns Over Variable Costs
				Cultivations	**Manure**		
1-1	Corn (110-day)	6/2	23,000	3	198-113-204	88	$249
1-2	Corn Silage	6/2	23,000	3	188-131-220	11 tons	$256
1-3	Red Clover	2nd year		-	-	2.1 tons	$132
					Plowdown		
2-1	Oats/Red Clover	3/16	48/10	-	-	78/0.9 tons	$115
2-2	Corn (95-day)	6/3	23,000	3	102-25-100	78	$232
2-3	Oats/Red Clover	3/16	48/10	-	-	76/1.1 tons	$137
				Herbicides	**Fertilizer**		
3-1	Soybeans	6/3	174,000	linuron metolachlor	-	42	$195
3-2	Corn (110-day)	5/11	23,000	cyanazine metolachlor	120-30-10	95	$188
3-3	Corn (110-day)	5/11	23,000	cyanazine metolachlor	120-30-10	102	$215

[1] Rate is pounds per acre for cereals and red clover, and seeds per acre for corn and soybeans.

Cash returns were highest for organic corn grain and silage in 1983.

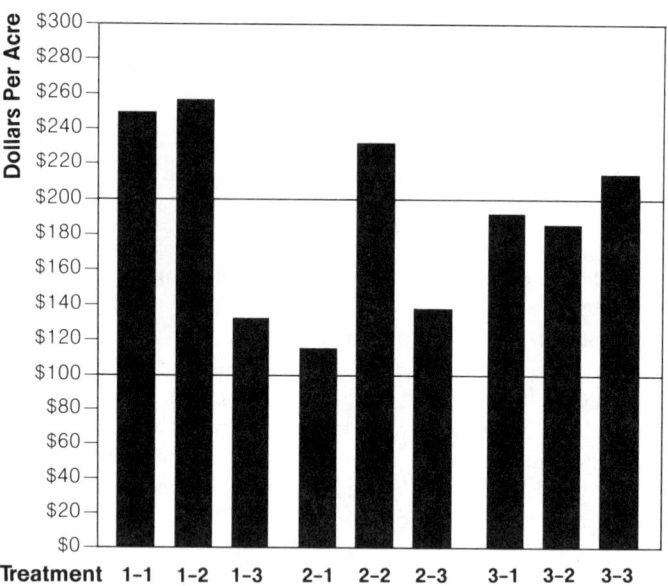

corn. This difference was not statistically significant. With higher plant populations, corn yields in 1-1 probably would have equalled or exceeded those in conventional corn.

Corn yields in treatment 2-2 were 20 percent less than conventional, and significantly different in the statistical analysis. The lower yields in this treatment could have been the result of different corn varieties, (treatment 2-2's corn had a relative maturity of only 95 days) or variable pollination conditions. With normal rainfall, conventional corn (treatments 3-2 and 3-3) would have been expected to yield in the 150-bushel range; treatment 1-1 about 135 bushels; and treatment 2-2 about 120 bushels, because it was a shorter-season hybrid. The drought didn't seem to affect the organic treatments as much as it did the conventional.

Leaf tissue analysis indicated that the drought in July and August prevented conventional corn (treatments 3-2 and 3-3) from using all of the fertilizer nitrogen that it received. Both conventional corn treatments were nitrogen-deficient, and had significantly lower leaf tissue nitrogen than the organic corn following two years of red clover (treatment 1-1). Organic corn silage following one year of red clover (treatment 2-2) had barely enough nitrogen, and organic corn following soybeans (treatment 1-2) was severely nitrogen-deficient. This deficiency was probably due to leaching of nitrogen in the manure that was applied the previous fall, the lack of moisture needed by soil microbes to release nitrogen from manure during the growing season, and less nitrogen carry-over from the soybeans compared with the clover plowdown.

Broadleaf weed biomass was significantly higher in all organic treatments. Grass weeds in treatment 1-1 were not significantly different from conventional treatments, indicating that the two previous years of red clover had helped control grass weeds.

Treatment 3-1, the only soybeans in 1983, yielded 42 bushels per acre, a good yield considering the drought.

The two spring oats/red clover treatments (2-1 following corn, and 2-3 following soybeans) yielded virtually the same: 78 and 76 bushels of grain, and 0.9 and 1.1 tons of hay per acre, respectively. Tissue analysis showed all nutrients were sufficient. Only one cutting of hay was made due to drought. Second-year clover in treatment 1-3 yielded 2.1 tons of hay per acre in two cuttings, the second of which was reduced by drought.

Economics—Despite lower yields for organic corn, cash returns were better than those for conventional corn.

Conversion Project Rotations

Organic Farming With Animals (System 1)

Treatment	1981	1982	1983	1984	1985
1-1	Spring Oats Red Clover Hay	Red Clover Hay	Manure Corn	Soybeans	Manure Corn Silage
1-2	Corn	Soybeans	Manure Corn Silage	Wheat Red Clover OS*	Red Clover Hay
1-3	Manure Corn Silage	Wheat Red Clover OS*	Red Clover Hay	Manure Corn	Soybeans

Organic Farming Without Animals (System 2)

2-1	Spring Oats Red Clover Hay	Corn	Spring Oats Red Clover Hay	Corn	Soybeans
2-2	Soybeans	Spring Oats Red Clover Hay	Corn	Wheat Hairy Vetch OS*	Corn
2-3	Corn	Soybeans	Spring Oats Red Clover Hay	Corn	Spring Oats Annual Legume OS*

Conventional Cash Grains (System 3—Control)

3-1	Corn	Corn	Soybeans	Corn	Soybeans
3-2	Soybeans	Corn	Corn	Soybeans	Corn
3-3	Corn	Soybeans	Corn	Corn	Soybeans

*Overseeding

Returns for organic corn following two years of red clover (treatment 1-1) were 23 percent higher than the average of the two conventional treatments, while the organic corn following soybeans and one year of oats/clover (treatment 2-2) returned 15 percent more. All corn yields were limited by drought, and again, the high input costs of conventional corn reduced cash returns. Corn silage (treatment 1-2) returned a hefty $256 per acre, due to low input costs and high market prices.

What if . . . ?

By the end of the '83 growing season, we'd begun asking ourselves a lot of "what if" questions. The one that came up most often was, "What if we'd taken the corn grain off as silage and sold it?" Not too many cash-grain farmers have that option, but we made the calculations anyway. We took the dry weight samples from the corn grain plots and estimated what the silage yields would have been, and their return per acre. But all that showed was that there was a definite advantage to farmers who can sell corn silage in a drought year.

What would be really helpful is knowing what our grain yields would have been if we wouldn't have had to replant in 1981. One way of estimating that is to convert the silage yield estimates back into grain. Agronomists estimate that 50 percent of the dry matter in silage is grain. This assumes sufficient moisture at grain formation, which we probably would have had if we'd planted earlier. If anything, this may overestimate grain yields a little.

While we're at it, we should also convert the silage from treatment 1-2 in 1983 and 1-3 in 1981 back into grain, so that their returns won't be artificially inflated by the seller's market for silage in those years. All our silage figures are adjusted to 65 percent moisture, so 1 ton of silage would

Beef manure and two years of red clover supplied plenty of nutrients to this 1983 corn. Yields were 10 percent lower than conventional corn, but cash returns were higher.

contain 700 pounds of dry matter, or about 350 pounds of grain. At 56 pounds of grain per bushel, that's 6.25 bushels of grain per ton of silage.

The reason for the better return from the lower yields in treatment 1-2 in 1983 was the better market price: $3.98 per bushel, compared with $2.61 per bushel in 1981.

The estimated average conventional corn yields in 1981—110 bushels, or 10 percent more than 1983's conventional corn yields—is realistic or perhaps a little generous.

What if . . . weather and soybean nodulation had been normal in '81 and '83? Chances are, the conventional systems would have earned slightly more than the organic.

Organic Farming With Animals

		Adjusted cash returns per acre		
Treatment	1981	1982	1983	3-yr. avg.
1-1	Oats/Red Clover $128	Red Clover $121	Corn Grain $249	$166
1-2	Corn Grain (Est.) $66	Soybeans $154	Corn Grain (Est.) $168	$129
1-3	Corn Grain (Est.) $79	Wheat/Red Clover $196	Red Clover $132	$136

Organic Farming Without Animals

2-1	Oats/Red Clover $139	Corn $111	Oats/Red Clover $115	$122
2-2	Soybeans (Est.) $175	Oats/Red Clover $49	Corn $232	$152
2-3	Corn Grain (Est.) $66	Soybeans $155	Oats/Red Clover $137	$119

Conventional Cash Grain

3-1	Corn (Est.) $94	Corn $187	Soybeans $195	$159
3-2	Soybeans (Est.) $169	Corn $163	Corn $188	$173
3-3	Corn (Est.) $104	Soybeans $120	Corn $215	$146

The county average for 1981 was 92 bushels, with moisture being the limiting factor. Since the Conversion Project field tends to be wet, the 20 percent better yield is not unexpected.

From our actual 1981 grain yields, however, we expected the organic treatments (1-2 and 2-3) to be more than 40 percent lower than the conventional (3-1 and 3-3) after adjusting the figures. So our estimated yields of 77 and 91 bushels per acre in the organic treatments look high. Assuming this 40-percent reduction is what actually would have happened, we'd estimate about 66 bushels per acre for these two treatments. Based on 1981 market prices, these treatments would return $66 per acre in 1981. The three-year average returns for treatment 1-2 would be $129 per acre; and for treatment 2-3, $119 per acre.

Measuring corn yields in weigh wagon. Yield figures from large-scale plots gave RRC scientists a more accurate picture of how each system performed.

Conversion Project Rotations

Organic Farming With Animals (System 1)

Treatment	1981	1982	1983	1984	1985
1-1	Spring Oats Red Clover Hay	Red Clover Hay	Manure Corn	Soybeans	Manure Corn Silage
1-2	Corn	Soybeans	Manure Corn Silage	Wheat Red Clover OS*	Red Clover Hay
1-3	Manure Corn Silage	Wheat Red Clover OS*	Red Clover Hay	Manure Corn	Soybeans

Organic Farming Without Animals (System 2)

Treatment	1981	1982	1983	1984	1985
2-1	Spring Oats Red Clover Hay	Corn	Spring Oats Red Clover Hay	Corn	Soybeans
2-2	Soybeans	Spring Oats Red Clover Hay	Corn	Wheat Hairy Vetch OS*	Corn
2-3	Corn	Soybeans	Spring Oats Red Clover Hay	Corn	Spring Oats Annual Legume OS*

Conventional Cash Grains (System 3—Control)

Treatment	1981	1982	1983	1984	1985
3-1	Corn	Corn	Soybeans	Corn	Soybeans
3-2	Soybeans	Corn	Corn	Soybeans	Corn
3-3	Corn	Soybeans	Corn	Corn	Soybeans

*Overseeding

Finally, we need to account for poor soybean nodulation in 1981. Since 1983's weather was similar to 1981's, we simply took the 1983 yield from treatment 3-1—42 bushels per acre—and coupled it with 1981 market prices. This gave us a three-year average return of $175 per acre for treatment 2-2 and $169 for treatment 3-2.

Although only ballpark estimates, these figures indicate what may have happened if rainfall and soybean nodulation were normal in '81 and '83. After three years, treatment 1-1 clearly stands out as the best *Organic Farming With Animals* rotation. Before we adjusted the yield estimates, treatments 1-2 and 1-3 appeared more profitable than they actually would have been under normal weather conditions.

Treatment 2-2 shows that soybeans may be a good crop to start an organic rotation without animals. Treatment 2-1 still lags behind economically, but should do better with two high-value crops—corn in 1984 and soybeans in 1985—yet to come in the five-year rotation.

Our number-crunching exercise has brought the three-year return for our best conventional rotation (3-3) up to slightly more than our best organic treatment (1-1). The unadjusted economic analysis implies that the organic rotations were clearly economically superior. Our adjusted and estimated analysis gives a more realistic picture.

After three years, cash returns on corn grain in the organic systems were finally higher than the conventional. As the organic rotation matures, nutrient cycling and improved weed control should make it even more profitable.

Keep in mind, also, that more of the costs in the organic treatment are for labor, not purchased inputs. Unless you hire a lot of custom work, that means more money in your pocket and less flowing off the farm.

1984

We were still analyzing data from the 1984 harvest as this book went to press. But so far, cash earnings from organic corn appear to be improving even further. In treatment 1-3, corn following two years of red clover hay yielded 139 bushels per acre, 8.5 percent less than the average from the two conventional plots. Yet the crop earned $301 per acre over operating costs. That's $21 more than treatment 3-1's 155-bushel corn.

As expected, cash earnings from treatment 2-1 are improving, too. Corn in that *Organic Farming Without Animals* system earned $247 per acre over operating costs, and increased the rotation's four-year average earnings to a respectable $153 per acre, better than two of the conventional systems.

Conventional and organic soybeans continued to produce almost identical yields: 52 and 53 bushels per acre, respectively. But the organic crop earned $27 per acre more, again due to much lower production costs.

Weather—June temperatures were slightly above average; otherwise, temperatures were normal during the growing season. Rainfall was 7 inches above average in May, and 5 inches above average in July.

Field Operations—The 2-year-old red clover in treatment 1-3 was so sparse that no biomass cuts were taken. Cattle manure was applied to corn in that treatment on April 27, and plowed the next day. Red clover in treatments 2-1 and 2-3 was plowed down on May 16.

In 1984, organic corn following two years of red clover hay outearned all other crops.

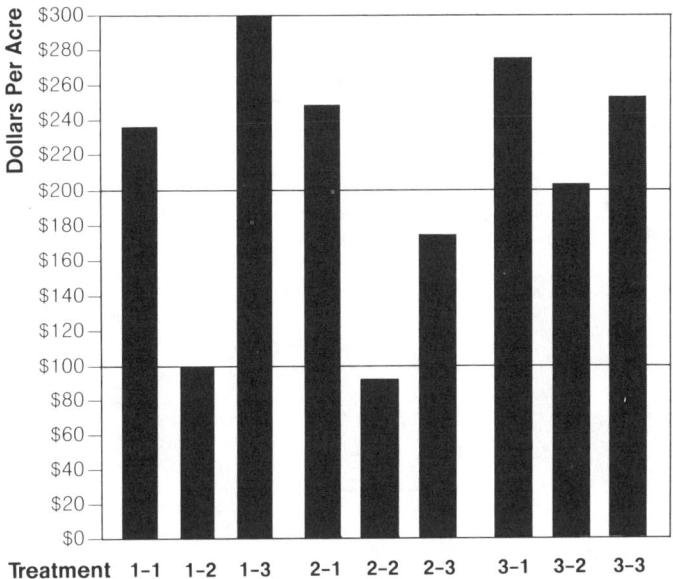

RRC scientist looks for changes in soil respiration when chemicals are withdrawn.

1984 Field Operations/*Lower production costs continue to make the organic corn more profitable.*

Treatment	Crop	Planting Date	Planting Rate[1]	Weed Control	N-P_2O_5-K_2O (lb/A)	Yield (bu/A)	Returns Over Variable Costs
				Cultivations	Manure		
1-1	Soybeans	5/26	175,000	2	-	53	$235
1-2	Wheat-Red Clover	10/4/83 3/27	120 15	-	-	37/1 ton	$103
1-3	Corn (110-day)	5/2	23,200	4	169-111-156	139	$301
					Plowdown		
2-1	Corn (110-day)	5/25	23,200	4	176-37-167	118	$247
2-2	Wheat-Hairy Vetch	10/17/83 9/6	120 40	-	-	35/1 ton	$96
2-3	Corn (95-day)	5/25	23,200	4	197-39-180	94	$179
				Herbicides	Fertilizer		
3-1	Corn (110-day)	5/2	23,200	dicamba[2] cyanazine metolachlor	120-30-20	155	$280
3-2	Soybeans	5/26	175,000	acifluorfen- sodium bentazon	-	52	$208
3-3	Corn (110-day)	5/2	23,200	dicamba[2] cyanazine metolachlor	120-30-20	149	$262

[1] Rate is pounds per acre for wheat, red clover and vetch, and seeds per acre for corn and soybeans.
[2] Spot-sprayed

Heavy rains in fall '83 and spring '84 reduced wheat yields in treatments 1-2 and 2-2. Red clover was cut and baled with wheat straw from July 23 to July 26. Excessive field traffic destroyed the Mammoth red clover and yellow sweet clover that had been overseeded into wheat in treatment 2-2. These plots were plowed, cultipacked and seeded with hairy vetch at a rate of 40 pounds per acre.

Starter fertilizer (100 pounds per acre of 10-30-20) was banded at corn planting in treatments 3-1 and 3-3. Cyanazine (Bladex) and metolachlor (Dual) herbicides were applied on May 7 at 1.4 quarts and 1.4 pints per acre, respectively. Dicamba (Banvel) was spot-sprayed on June 15 where thistles and bindweed were particularly heavy. The corn was sidedressed on June 21 with 110 pounds of nitrogen (34-0-0) per acre.

Soybeans were sprayed with 1 pint per acre each of acifluoren-sodium (Blazer) and bentazon (Basagran).

Finally, corn in treatment 2-3 dried down slowly, so we decided to plant spring oats instead of winter wheat in those plots in 1985.

Conversion Project Rotations

Organic Farming With Animals (System 1)

Treatment	1981	1982	1983	1984	1985
1-1	Spring Oats Red Clover Hay	Red Clover Hay	Manure Corn	Soybeans	Manure Corn Silage
1-2	Corn	Soybeans	Manure Corn Silage	Wheat Red Clover OS*	Red Clover Hay
1-3	Manure Corn Silage	Wheat Red Clover OS*	Red Clover Hay	Manure Corn	Soybeans

Organic Farming Without Animals (System 2)

Treatment	1981	1982	1983	1984	1985
2-1	Spring Oats Red Clover Hay	Corn	Spring Oats Red Clover Hay	Corn	Soybeans
2-2	Soybeans	Spring Oats Red Clover Hay	Corn	Wheat Hairy Vetch OS*	Corn
2-3	Corn	Soybeans	Spring Oats Red Clover Hay	Corn	Spring Oats Annual Legume OS*

Conventional Cash Grains (System 3—Control)

Treatment	1981	1982	1983	1984	1985
3-1	Corn	Corn	Soybeans	Corn	Soybeans
3-2	Soybeans	Corn	Corn	Soybeans	Corn
3-3	Corn	Soybeans	Corn	Corn	Soybeans

*Overseeding

Despite poor wheat yields, 1984 was the most profitable year yet for the Conversion Project. Treatments 1-1 and 1-3 show that corn is most profitable when grown after two years of red clover. And organic soybeans have outearned conventional soybeans every year, even though yields are virtually the same.

Overall, the organic rotations that start with oats/red clover had the highest cash returns for the first four years.

A three-year (1981-83) technical summary of the Rodale Research Center's Conversion Project is available for $2 from: Rodale Research Center, R.D. 1, Box 323, Kutztown, Pa. 19530. Ask for report #RRC/AG-84/1.

HYBRID HY-JINKS OR HOW NOT TO DO ON-FARM RESEARCH

(THE SCENE: A crossroads coffee shop somewhere in the Corn Belt.)
FARMER JONES: I'm really sold on that new SUPERYIELD hybrid corn I tried this year. Best yields I've ever had. I've got 170-bushel corn out there just waiting to be harvested. That seed was worth every extra dollar I paid for it. Guess I'll go with it on the whole farm next year. Probably make a killing.
FARMER BONES: Probably. Tell me more about your experiment.
FARMER JONES: Well, I rented that 40-acre field down along the river from my cousin Fred. Probably have to make 100 trips in the semi to get all that grain back up here on the ridge.
FARMER BONES: Fred's sold off the last of the herd hasn't he?
FARMER JONES: That's right, he's getting out for good. Too bad, too. This field was no nice. Second-year alfalfa he'd limed and manured. He didn't even get a chance to take a cutting off last summer.
FARMER BONES: Good soil down there?
FARMER JONES: Black as coal, not stony like here on the ridge. Plus high water last spring must have left me another inch of soil, to boot.
FARMER BONES: With all that water you must not have been able to get into the field 'til late.
FARMER JONES: Heavens, no! Water dropped down and with that good southern exposure it dried up in a hurry. It was the first field I planted. That SUPERYIELD's 110-day corn, not like the 95-day I use up here on the ridge.
FARMER BONES: Wasn't the best of growing seasons, either. Awful dry this summer.
FARMER JONES: I'll be lucky to get 100-bushel corn on the ridge. But down by the river, that field just soaked up those little rains we had. Boy, I'm sold on that SUPERYIELD!
FARMER BONES: You're only getting 100-bushel corn on that ridge this year? I'd have thought you'd get at least 120, even with the drought.
FARMER JONES: I'd probably have gotten a lot more than that if I'd gone with that SUPERYIELD up there. I had lots of disease and insect problems up there. County agent says I should diversify and at least work some beans into rotation to break up the corn.
FARMER BONES: Sounds like a good idea.
FARMER JONES: Well, 100-bushel corn's just barely gonna pay for itself. But based on my experiment down on Fred's field, I can make up for it next year if I go with SUPERYIELD. Probably make a killing.
FARMER BONES: Probably.

Epilogue:

Farmer Jones did make a killing, but his business is what died. Needless to say, SUPERYIELD did not do as well on the ridge—with the different soils, cropping history, and microclimate—as it did on the bottomland. While no one would be foolish enough to draw the conclusions Farmer Jones did from his so-called experiment—that the new hybrid was responsible for better yields—at least he was trying out a new idea on his own.

As you take charge of your own operation, on-farm experiments can be an increasingly important way to fine-tune your management practices.

Try It . . . MAYBE You'll Like It

As a farmer, you are constantly bombarded with information. The fertilizer dealer tells you his program will increase your yields. The pesticide dealer has just the thing to take care of your "weed problem." You see an ad for a new corn variety that's resistant to a particular insect you've been having problems with. Extension agents have their own views on how you should run your farm. Now here we come suggesting you reduce your chemical use. Who are you going to believe? What are you going to do?

While most of the information you receive is backed up by some experimental data, your situation is unique. You may also have different questions and concerns from those of people conducting the experiments. Herbicide manufacturers want to know how *effectively* their product eliminates weeds. You're probably more concerned with how *cheaply* it eliminates them, and may even tolerate a few weeds if yields aren't reduced. Your best source of information may just be your own farm.

"Try it out on a few acres, first. Then see what works best for you." Our agronomy staff at the Rodale Research Center has given this advice to farmers countless times. But a researcher's idea of trying it on a few acres, and a farmer's idea, are probably a lot different. Their results will be, too.

Why? Mainly because the time that each has to plan and conduct experiments differs. To a researcher, an experiment is a tightly controlled, carefully monitored project. Researchers are paid to design randomized and replicated experiments that yield statistically valid, scientifically defensible results.

Farmers are paid for the crops they grow. Although we know some farmers, most without formal training, who are doing some pretty sophisticated on-farm research, it's unlikely that most have the time to develop a research project as carefully as a researcher. However, with a little thought and effort you can do research that will prove to yourself what works and what doesn't, and quickly pay for the time you've invested. Here are some things to consider as you get started:

Gather Information—Before you start planning your experiment, find out as much as you can about the subject. Read as much as you can, talk to your Extention agent and other agricultural professionals, and share your ideas with neighbors. Their thoughts and advice can help clarify in your own mind what your experiment is all about.

Keep Control—The purpose of your experiment will probably be to compare the results of two management practices. For example, you can compare fertilizer rates, crop varieties, planting densities, or weed management strategies. These are called the treatment variables. Keep the number of variables small. Include more than one or two, and you'll have a hard time figuring out which ones are responsible for different results. Test different fertilizer rates while keeping all the other management practices the same. Or, test different weed management strategies with the same variety and fertilizer rates.

The Unknowns—There are three major variables over which you can have no control: weather, soil, and the genetic diversity of the plants you grow. But there are things you can do to make sure these variables aren't affecting your results.

For starters, keep in mind that weather is more than the daily maximum and minimum temperatures and total precipitation. It may take years to develop the intuition necessary to relate crop yields to weather conditions, but *a good maximum/minimum thermometer and a rain gauge* can be indispensable tools to help you develop and refine these skills. So if you don't already do so, **begin keeping accurate weather records** for your farm. You can use local weather station data, but the microclimate of your farm still may be substantially different. *Climatological Data of the United States Weather Bureau* will provide you with weather data from the nearest U.S. Weather Bureau Station. Look for it in your local library or order it from: Publications, Climatic Center, Federal Building, Asheville, N.C. 28801.

One way of taking variable weather conditions into account is to run the same experiment over several years. Does one treatment do better every year? If so, weather probably isn't having much effect on your results. Does one do better during wet seasons and another during dry? That's interesting to know, but not terribly helpful, since there's no way predict rainfall and moisture a season ahead.

Also, make sure you know the natural classifications of the soils in your fields, and their cropping histories. Take soil samples before experimenting so that you know nutrient levels at the start.

Setting up experiments on soils with uniform soil type and fertility is desirable, but not necessary. There are two techniques you can use to take these differences into account. For example, where there is a soil gradient, and soil fertility generally decreases as you go up the slope, be sure that your treatments run with the gradient. This way, each treatment has equal amounts of all soil types. (We are not recommending you plow and plant up and down the slope.)

Better yet, if you have a lot of variability in your soils, repeat your treatments in pairs on different soil types. This will help you see if the treatments perform equally well on a wide range of soils. One treatment may perform better on marginal lands, while another works best on more fertile soils.

Finally, even though the seed you use has been bred to produce uniform plants, each plant will grow and yield a little differently. Harvest and measure a large enough sample to be sure your field is representative. Cross-pollinated crops like corn and rye are more variable than self-pollinated crops like beans, wheat and barley. Therefore, you'll need a larger sample for the cross-pollinated types.

Ideal Plot Size—The ideal plot is big enough to take into account plant variability and small enough to provide relatively uniform soil conditions and yields consistent results. Scientists are able to get good results on small plots—.001 acres for vegetables, .025 acres for field crops—because they have the tools and the time to measure accurately. In determining how large an area to devote to your experiment, you are probably limited most by the accuracy with which you can measure yields. The larger the area you measure, the more accurate your results will be. Just be sure to take into account differences in soils when you do your experiment in a large area.

Border Effect—Everyone knows that those spindly corn plants on the roadside edge of a field aren't representative of the field as a whole. If you harvested just the border and tried to predict your yields, your prediction would be way too low.

Don't let this border effect mislead you in your on-farm experiments, either. If you run your experiments in long strips or along the edge of a field, only measure the center rows. They are most truly representative of the effect of the treatment. Outside rows might be affected by fertilizer and chemical drift, weed competition, unequal tillage or cultivation, and a whole host of confounding factors. Another alternative is to plant a large square area, where the border effect will be negligible on the overall yield.

Replications—If you repeat (replicate) the treatments in different plots placed around the farm, you can be more sure your results weren't a one-time, chance occurrence. Three to six replications is the standard for most agronomy experiments. If a treatment works great on some plots, and fails miserably on others, you may need to repeat that treatment more often or settle for less certainty in your data.

Keep Track Of Economics—It's sometimes difficult to take everything into account when doing an economic analysis. Purchased inputs shouldn't pose too much of a problem if you keep good records of costs and know your application and seeding rates. Field operations may be more difficult to estimate, since labor, fuel, depreciation, and other hard-to-measure costs are involved. For comparison purposes in the RRC Conversion Project, we used custom rate guides (available from your state department of agriculture or county agent), because they include the cost of all these variables. The guide assumes you hire out all your operations, so it gives you a fairly valid comparison of field operations for different management practices. If anything, it will make the treatment with the most operations look slightly more expensive than it really is.

Make Observations—Don't wait and just measure yields at the end. Get down off your rig and take a closer look at your plants and soil. How does that corn look at lay-by? Can you see differences in height, color and general vigor? Keep careful records of your observations so you can refer to them later. With soil fertility questions, it's helpful to look for nutrient deficiencies in your plants. A good resource for this is the book "Hunger Signs in Crops," edited by H.B. Sprague (1964, David McKey Co., New York). Analysis of leaf tissue samples is another good way to check for nutrient deficiencies and predict nutrients that will be available to the following crop. Your county agent can recommend a laboratory and supply sampling and drying directions. Cost is usually from $10 to $20 per sample.

Try To Figure Out Why—Even though your experimental results and other observations may not tell you why some things work and others don't, always look for explanations. The whys sometimes hide themselves. Soil fertility, moisture and weed competition may be closely interrelated and you may be able to discern patterns in your fields that explain what works for you, but are difficult to see experimentally.

Peer Review—Scientists put their research plans and results out for all to see and comment on. You should do the same. Go to the coffee shop, Grange, or wherever, and get your friends and neighbors excited about your research. You'll get them thinking about what you're doing, and maybe they'll even be curious enough to try a few test plots on their own.

A Hypothetical Experiment— Herbicide vs. Cultivation

So you'd like to reduce your herbicide use. Even though you're rotating herbicides, weed control gets worse each year. In the meantime, your chemical bills are going through the ceiling. You'd like to find out if cultivation will fit your style, control weeds and maintain high yields. How would you devise an experiment to answer your questions?

First, state your question clearly. 'Will cultivation give me cash returns equal to or better than weed control with herbicides?' While this is your primary concern, you can design the experiment to answer other questions, as well.

You decide to make the test on that 40-acre field where you have yet to see any significant variations: no wet spots, rock outcrops, or the like. It's not your best field, or your worst, but is fairly typical of your farm as a whole. The field is rectangular, running east and west, and there's a slight north-south drop in elevation. Sure enough, when you check the soil map, you find the lower quarter of the field is a different soil type. You had originally planned two plots running east to west. But with the variation in soil types, this arrangement might favor one treatment over the other. Instead, you decide to go with two square plots with a large buffer strip running north and south through the middle of the field. This way, both treatments will have equal areas of the two different soil types. The buffer strip will also prevent herbicide drift and give you lots of space to turn around. Consider the lost cropping area an investment in information.

As you're making these plans for the coming season, start a journal of observations. Before the snow flies, tour the field and see what kinds of weeds have gone to seed, how much weed pressure there's going to be, and if there appears to be any difference between the two plots you've planned. If it looks like weed pressure will be unequal, you may want to lay out a more complicated set of plots so that each treatment will have an equal share of weedy areas.

A quick check behind the equipment shed reveals that the old cultivator deposited there years ago is not exactly state-of-the-art. A visit to the local equipment dealer proves fruitful, however. You explain your intentions, and he gives you a cut-rate lease on his best new cultivator. He'll even let you keep it at your place, so you can cultivate at the right time without the hassle of transporting it— provided, of course, that you tell him how the experiment works out, so he can convince some of your neighbors to buy the new line.

As spring approaches, think through the experiment and start your record-keeping. Your management practices for each field should be identical, except for herbicides and cultivations.

Continue your observations and record-keeping throughout the season. Some information may not answer your basic question, but it may be useful in the future. For example: Are your most persistent weeds in the cultivated field broadleaf or grass? Are they between rows or in rows? Were corn populations reduced by cultivation? Did weather allow you to cultivate when you needed to? Would you have been able to cultivate the rest of your farm? Did weather favor weed germination and growth? Always keep questions like these in mind.

Assess weed pressure between the two treatments at peak weed growth. There are scientific sampling methods you can use (contact your county agent for details), but chances are your observations will provide enough information. There are probably significantly more weeds in the cultivated field that will be visible to the eye, though not necessarily enough to reduce yields.

Now comes harvest, and a chance to see if there are any actual yield differences between your two weed control techniques.

Once yields are known, calculate total returns over variable costs for the two treatments. The answer to your original question will be obvious. But over the winter, as you think through your results, you'll probably find out that you've generated more questions than answers, both about the *reasons* for your results and about the *methods* used to generate them.

That 40-acre field may become a permanent on-farm laboratory, where each year you conduct more and more sophisticated experiments on smaller plots. The money you save as you apply the results of your own research to the rest of your farm may just help you swing that cultivator purchase, too!

When In Doubt, Ask A Farmer

Research results—both ours and your own—can be a good way to test cost-cutting strategies. But when it comes to anticipating problems and finding workable solutions, nothing beats the advice and experience of other farmers. For example, how many times have you wondered what *really* happens to a farm's income when chemicals are reduced or eliminated? Or what happens to yields. Or whether weeds, nutrients or some other factors are as much of a problem as everyone says.

We've wondered the same things for a long time. So in spring '84, we posed these and other questions to the largest group of farmers we know of who are cutting back on chemical use: readers of *The New Farm* magazine.

The questionnaire at the end of this chapter first appeared in the March/April '84 issue of the magazine, and has been reprinted several times since then. By October '84, some 300 farmers from throughout the United States and Canada—and even a few from Europe—had told us why they wanted to use fewer chemicals, what obstacles they'd encountered when they began doing so, and how they overcame them. Many included handwritten notes or letters describing their operations in more detail, and offering advice for other farmers.

The average farmer we surveyed cultivates from 100 to 400 acres, and raises corn, soybeans, hay or small grains, and at least one type of livestock. Most began reducing chemical use in the late '70s or early '80s.

Questionnaires were still arriving when this book went to press (if you haven't filled one out, please do so), and we'll continue tabulating and updating them as this new farmer-information network grows. Meantime, what we've learned so far confirms many theories about reducing chemical use, disproves others, and even raises a number of new ones. First, a profile of the farms and farmers we heard from.

Farm sizes range from just a few *tillable* acres to more than 1,800. Actually, about half of the farms are from 100 to 400 acres; 32 percent are smaller than 100 acres, and 20 percent are larger than 400, with 7 percent exceeding 750 acres. Corn and soybeans are grown on 60 percent of the farms, and hay and small grains are grown on 54 percent. Other cash crops include straw (26 percent of the farms), fruits and vegetables (22 percent), and assorted specialty crops like dried beans, seed crops and tobacco (fewer than 5 percent).

All but 50 of the farmers raise at least one type of livestock. About half raise beef cattle, one-third raise hogs, one-fourth raise dairy cows, and one-fifth raise poultry. Sheep, horses, goats and other livestock are raised by fewer than 15 percent of the farmers.

Finally, of the 247 farmers who said they'd already begun reducing chemicals, a little more than two-thirds had done so sometime between the late '70s and early '80s. Only a handful are long-time organic farmers.

Nearly three-fourths quit using chemicals for financial, health or environmental reasons–usually all three.

'Cheaper, Safer, Healthier'

"The reason we want to do this is not only because it is cheaper, but also because it is safer," wrote Richard J. Teubel, who raises dairy cattle, grains and hay on 185 acres in Bloomfield, Iowa. Teubel's remarks almost perfectly describe the feelings of the other farmers we surveyed. Many gave multiple reasons for cutting back on chemicals, but by far the most popular were "Cut production costs," "Environmental concerns," and "Personal/family health and safety," each of which was cited by more than three-fourths of the respondents.

Many farmers carried Teubel's feelings one step further, and offered some practical, albeit fairly general advice. "Fit your system to your climate and soil," said Texan Mackie Allgood, who pastures cattle and sheep, and raises small grains on 200 acres. "Be flexible. Enjoy farming enough to *work*, or do something else. All good farming is labor-intensive."

> *More than 75 percent said yields and income either stayed the same or increased when they cut back on chemicals.*

"You should enjoy yourself and what you're doing," agreed Norm Zeman. "It's great. It's a way of life." Zeman, one of the few long-time organic farmers who answered the questionnaire, raises beef cattle and hogs, and rotates his 120 acres in Ottosen, Iowa, to corn, beans, oats and hay. His letter read almost like a Ten Commandments for farmers: "There is no right way to farm . . . Whatever works for you . . . Don't expect purchased products to perform miracles." And finally, "(Reducing chemicals) won't be easy. You might be talked about!"

Judging from our survey results, though, the only gossip these farmers have had to endure was pretty favorable. Nearly three-fourths of them said crop yields either stayed the same or increased when they started using fewer chemicals, while just 37 percent noticed yield declines in some crops. (The answers add up to more than 100 percent because some farmers experienced declines and increases.) "This is just the second year we have partially farmed organically," said Pennsylvanian Robert Smith, who raises beef and dairy cattle on 468 acres. "Corn after clover/alfalfa received no fertilizer; 10 tons of manure per acre was applied. We are still using a herbicide — no insecticide, though — but would like to change to cultivating in future years. Yields were about the same on this land as on other corn where chemical fertilizer was used according to soil test."

The farmers who did report yield declines generally felt it took from two to three years for them to recover, a conclusion shared by scientists at the Rodale Research Center and at other research facilities which have studied what happens when chemicals are withdrawn from a cropping system. A few survey farmers said yields rebounded in just one year, while others said yields still hadn't recovered.

A Better Bottom Line

The effect that management changes had on net farm income was even more pronounced. A whopping 88 percent of the 213 farmers who responded to this item on the questionnaire said their net income either stayed the same or *increased* when they began farming with fewer purchased inputs. Just 13 percent said net income declined.

Having livestock didn't seem to offer any particular advantage to yields or income. When compared with the respondents as a whole, slightly more of the farmers with livestock said they noticed yield declines when they began reducing chemicals. But slightly more of them noticed yield increases, too. Likewise, income increases and declines were reported by an almost identical percentage of respondents, both when livestock farmers were included in the total, and when they were examined separately.

R. E. Corcoran of Elvins, Mo., credited a flexible crop plan with maintaining cash-flow on his 140-acre, mixed crop-livestock farm. "For several years, inputs were reasonable and I did well. But I have always kept track of *costs to plant*. And as prices changed, it became obvious that it would be profitable to go back to grass, and buy corn again.

"In this area, hay is more profitable, provided you have the equipment and some available laborers," Corcoran explained.

Fifty-three farmers said their net income increased even when yields stayed the same, and 19 said net income increased when yields declined. In fact, only 27 farmers reported declines in net income when yields fell or stayed the same.

> *Weed control is a problem for 71 percent, and nutrient deficiencies for 41 percent. Few, if any, farmers have difficulty with insects and plant diseases.*

Some 112 farmers offered a reason — usually several reasons — for yield reductions when they cut back on chemicals. (The rest said yields didn't decline, or gave no answer at all.) Weeds caused yield reductions on 71 percent of the farms, while plant nutrients reduced production on 41 percent. A few farmers said poor weather, a lack of proper equipment or a lack of know-how were their main stumblingblocks. Only 3.5 percent of the farmers — four, in all — said insects were responsible for lower yields when chemicals were withdrawn. None cited plant diseases.

The fact that weeds were the most frequently mentioned problem probably comes as no surprise. What is intriguing, though, is the success with which these farmers have been able to control weeds. Seven farmers took the time to explain their weed management strategies, and all but one reported higher net incomes now that they're relying less on chemical weed controls.

Cover Crops, Tillage Replace Herbicides

"I've found out that if you plant rye in cornfields after harvest it helps control weeds and grasses," wrote Illinois dairyman David L. Marcure, who said both his yields and net income improved when he quit using chemicals. "I also plant rye in the fall in cornfields that are going to be in oats the next year. Then, in the spring, I just disk and harrow, (before planting). There will still be some rye . . . but not enough to hurt anything. You have a much cleaner stand of oats with hardly any weeds or grass."

John Myer also said his net income has increased since he began reducing chemicals around 1980. Myer raises sheep, and dairy and beef cattle on 300 acres in Ovid,

N.Y., and feels herbicides simply aren't cost-effective in small grain and hay crops. "In 1982, I planted 14 acres of alfalfa on April 18," he wrote. "After the alfalfa came up, I sprayed a 30-foot-wide strip with 'Premerge' herbicide as a check. It was nearly impossible to locate the check strip, except for a heavy growth of foxtail in the wheel tracks of the tractor and sprayer. The thick early growth of alfalfa . . . shaded out the weeds.

More than half bought no new equipment when they began cutting back on chemicals. Those who did bought cultivators, rotary hoes, chisel plows and haying equipment.

"The same is true for oats and spring wheat," he continued. "I haven't used herbicides on small grains or alfalfa seedlings in three years, with *no* reduction in yield. The key is to remain flexible in your crop rotation. Plant a crop that will do well in that field *that* year. If a field is too wet to work early, plant a warm-weather crop."

Horse farmer Larry Olson, who also raises hay and mixed grains on 229 acres in Granite Falls, Minn., seemed to take a page directly from Dick and Sharon Thompson's weed management textbook: "The less tillage I did, the less problem I had with weeds," Olson wrote. "That's why I've moved to ridge-tillage."

The Thompsons, contributing editors to The New Farm, made a similar conclusion about tillage several years ago, when they began experimenting with ridge-planted corn and soybeans on their 300-acre Iowa farm. "We have found that the *deeper* you till and the *more* you till, the *more weeds* you'll have," say the Thompsons in their popular slide show.

Olson began cutting back on chemicals in the early '80s, and wrote that the effect on yields has varied: Corn has declined, small grains remain unchanged, and soybeans have increased. "If I were to start again, I would divide my farm into four fields and start with a two-year sweet clover rotation (on half my crops). I'd mow the clover the first year to control weeds, and plow it the second year for green manure.

Clovers and alfalfa are by far the most popular sources of plant nutrients.

"After building the soil with sweet clover, I would begin ridge-planting row crops on that half of the farm, leaving the ridges in for about 10 years, rotating corn and soybeans and adding vetch to corn after last cultivation. On the other half, I would rotate small grains and alfalfa, with sweet clover in the small grain each year."

Illinois farmer Mike Raftery uses a rye cover crop to help control weeds on his 120-acre hog and beef farm. He broadcasts about 90 pounds of rye per acre right after soybean harvest, then chops it and disks it in spring. "My experiences with rye show you need not make any special trips across the field when planting," he noted. "Also, I have started using the ridge-till system by Buffalo. I feel it gives me all the benefits of a minimum-till situation and the advantage of good mechanical weed control." Yields of corn, soybeans, small grains and alfalfa hay have stayed the same on Raftery's farm, while his net income has increased. What's more, he added, "the tilth of the soil is really coming back."

Not all of the farmers in our survey had to invest heavily in new equipment, though. In fact, more than half said they bought no new equipment at all when they decided to cut back on chemicals. Of the slightly more than 100 who said they did have to visit their machinery dealers, some 35 percent bought cultivators, 22 percent bought rotary hoes, 20 percent bought chisel plows, 15 percent bought haying equipment and 12 percent bought manure spreaders. Eight farmers bought some type of ridge-planting equipment, while a few others bought grain drills, disks, loaders, livestock or other items.

Obviously, mechanical weed control was pretty high on their list of priorities, though the means for achieving it varied widely. For example, Richard Laska, of Winona, Minn., swears by his rotary hoe. "A rotary hoe is a must if you don't use herbicides," he wrote. Yet David C. McCoy, who raises beef and dairy cattle on 360 acres in Ohio, sounded like he was about ready to sell his. "I need to find a better way to control weeds *in the rows* of my corn. My rotary hoe just isn't doing a good job."

Perhaps Texan Rick Watzl, who began reducing chemicals on his 700-acre cotton and grain sorghum farm in the early '80s, best captured the respondents' feelings: "Apply herbicides only when needed and not as a habit."

Nitrogen is the biggest nutrient worry for 62 percent of the farmers, while P and K each are a problem for 43 percent.

Green Manures Supply N

Nearly 200 farmers reported that nutrient deficiencies emerged when they cut back on chemicals. Many had problems with more than one nutrient, but, again, the deficiencies often weren't enough to affect yields. Lime, sulfur and magnesium were mentioned by a few farmers, and P and K were each cited by 43 percent.

Nitrogen was by far the biggest nutrient problem; it was listed by 62 percent of the farmers. That's probably why so many of the survey respondents include leguminous cover crops in their rotations.

Overall, about three-fourths of the farmers who wrote us said they use at least one green manure or cover crop. Some 58 percent still use some synthetic fertilizers, while an almost identical number use fresh animal manure to supply at least a portion of their crop nutrient needs. Organic fertilizers and composted animal manures are each used on about one-fourth of the farms, and sludge and other organic by-products are used on fewer than 7 percent.

Since green manures and cover crops are the most popular nutrient sources, which ones seem most effective? Clearly, it's legumes—clovers and alfalfa, to be exact. Of the 259 farmers who listed the green manures and cover crops they use, 61 percent said they rely on clovers, and 53 percent rely on alfalfa.

Small grains are grown as green manures or cover crops on about one-third of the farms, and ryegrass, vetch and buckwheat are grown on a few. Other farmers mentioned brassicas, fescue and mixed grasses.

Your Region At A Glance

Here's a look at some of the regional trends that are starting to emerge from our survey. Scanty responses from many parts of the country—the Southern Plains and the Northwest, for example—make it difficult to predict accurately the types of problems that farmers in these areas can expect, and their possible solutions. For example, you'll notice that a surprising percentage of farmers said corn did fine during their first year of reduced chemical use. Few farmers and researchers would support that view, yet it apparently has worked for a few growers. The only way to know why, though, is to go back to each individual questionnaire and study the entire operation from the ground up.

In fact, none of this information can tell you exactly how to reduce chemicals on your farm, or what will happen when you do. Its purpose is to show what has happened to some farmers, and what they have done about it.

Virtually all farmers grow a small grain or legume hay. And while 38 percent said yields dropped when they cut back on chemicals, 87 percent said income stayed the same or increased.

best, while 27 percent said soybeans did. Nearly one-fourth of the farmers also said corn yielded well, but considering the large number who use leguminous green manures, this may not come as such a surprise.

- **What happened to yields when chemicals were reduced/eliminated?:** Only 90 farmers answered this question. Most—about 60 percent—said yields didn't change, while 38 percent said yields declined, taking up to five years to return to normal. Roughly 10 percent of the farmers said yields of some crops increased.
- **Why did yields decline?:** Forty-eight farmers gave reasons, with weeds being cited by about 72 percent of them. Some 36 percent of the farmers mentioned plant nutrients as a cause of yield reductions, while fewer than 15 percent said a lack of proper equipment, or "other reasons" (usually rainfall) were a problem. Two farmers noted that insects caused yield declines.
- **Which plant nutrients were a problem?:** Sixty-eight percent of the 65 farmers who answered this question said nitrogen was their biggest nutrient problem. P and K each were a problem on about one-fifth of the farms, while lime and sulfur were cited by a handful of farmers.
- **What happened to net income?:** Of the eighty-seven farmers who responded, 45 percent said their net income stayed the same when they began cutting back on chemicals, and 43 percent said it increased. Only 11 farmers—13 percent of the respondents—reported declines in net income.

Midwest

Responses came from 114 farmers in Ohio, Illinois, Indiana, Iowa, Nebraska, Kansas and Missouri. Farm sizes range from less than an acre to 1,800 acres. Most farms are in the 100- to 500-acre range; 26 farms have fewer than 100 tillable acres, nine have more than 500 acres. Eighty-six of the farmers grow corn, 73 grow soybeans, and 101 grow a small grain or legume hay. Only seven of these farmers raise row crops exclusively.

- **Cover crops/green manures used:** One hundred farmers said they use at least one type. Alfalfa is grown by 61 percent, clover by 56 percent, and small grains by 41 percent. Other green manures and cover crops include rye (grown by 12 percent of the farmers), hairy vetch (8 percent), grasses (3 percent), and buckwheat.
- **Crop that did best during first year chemicals were reduced:** Seventy-six farmers answered this question (many gave more than one answer). Some 40 percent said hay did

North-Central

Responses came from 64 farmers in Michigan, Wisconsin and Minnesota, whose farm sizes range from seven to 1,300 acres. Thirty-nine of the farms are from 100 to 300 tillable acres; 18 are smaller than 100 acres, and seven are larger than 300 acres. Corn is grown on about 85 percent of these farms, and either forage legumes (mostly alfalfa) or small grains (wheat, oats and barley) are grown on an equal number. Only a little more than one-third of the farmers raise soybeans, and just two farmers raise vegetables. Four farmers raise row crops exclusively.

- **Cover crops/green manures used:** Of the 53 farmers who said they use cover crops or green manures, nearly three-fourths use alfalfa, and 69 percent use clovers. Small grains are far less popular in this region: Only 26 percent of the farmers use them as green manures or cover crops. Likewise, fewer than 12 percent of the farmers use rye, hairy vetch, brassicas or buckwheat.

- **Crop that did best during first year chemicals were reduced:** Forty-five farmers answered this question, and about half said hay was the easiest crop for them to grow without chemicals. The next easiest? Corn, which nearly one-third of the farmers said did just fine when they cut back on chemicals. The fact that nitrogen was the most often reported nutrient deficiency in the North-central area (and in most other regions) still makes corn a risky crop to grow during the first year of chemical reduction, though. Finally, soybeans and small grains were each successful on 20 percent of the farms.

About 40 percent of the farmers—more than in any other region—said yields declined when they reduced chemical use. Despite that, 91 percent said income stayed the same or increased.

- **What happened to yields when chemicals were reduced/eliminated?:** Twelve farmers didn't say, but the majority—60 percent—noticed no change in yields. The North-central states led all regions in the percentage of farmers who said yields declined: 40 percent. Meantime, only 12 percent of the region's farmers said crop yields increased when chemicals were reduced or eliminated.
- **Why did yields decline?:** Again, 12 farmers didn't answer the question, while an additional 26 said yields didn't decline. The vast majority (95 percent) of those who cited a reason said weeds were their biggest problem, followed by plant nutrients (40 percent) and a lack of proper equipment (25 percent).
- **Which plant nutrients were a problem?:** Forty-nine farmers reported nutrient deficiencies, with half listing nitrogen. Some 38 percent of the farmers cited problems with potassium, and only 20 percent cited phosphorus. Lime and micronutrients were again mentioned by just a few farmers.
- **What happened to net income?:** Of the 55 farmers who answered this question, only four said net income declined when they reduced or eliminated chemicals. About half said net income stayed the same, and 40 percent said it went up.

Green manures and cover crops are more popular in the Northeast than in any other region. That may explain why several farmers said they had little trouble cutting back on chemicals in corn.

Northeast

Responses came from 32 farmers in Maine, Massachusetts, New York, Pennsylvania and New Jersey. Most farm from 50 to 200 tillable acres. Ten of the farms are smaller than 50 acres, and four are larger than 200. Corn, small grains and forages are the most frequently mentioned crops, though some of these farmers grow fruits and vegetables.

- **Cover crops/green manures used:** Three farmers said they don't use them. Small grains and clovers—each used on 16 of these farms—are the most popular green manures/cover crops. Alfalfa is used on 10 of the farms, while hairy vetch, rye, buckwheat and grasses are each used on one or two.
- **Crop that did best during first year chemicals were reduced:** Only 16 farmers answered this question. Half said hay did best the first year, and 10 percent said small grains did, too. One-third of the farmers also reported success with corn.
- **What happened to yields when chemicals were reduced/eliminated?:** Of the 23 people who answered this question, 16 said yields stayed the same, and four said yields of at least some crops increased. Six farmers noted declines in some crop yields, and indicated that it took two to three years for them to return to normal.
- **Why did yields decline?:** A number of farmers gave multiple answers. But generally, weeds were again the most frequently mentioned reason for yield reductions. Plant nutrient deficiencies, insects and other reasons were also cited, though only by one or two farmers each.
- **Which plant nutrients were a problem?:** Twenty farmers listed nutrient deficiencies, with nitrogen, phosphorus and potassium being cited by 11, nine and two farmers, respectively.
- **What happened to net income?:** Of the 20 farmers who answered this question, most said their net income either stayed the same (eight) or increased (seven), while five said net income declined.

South

Responses came from 18 farmers in Tennessee, Arkansas and Kentucky. Most farm from 100 to 300 tillable acres. Seven of the farms are smaller than 100 acres, and three are larger than 300. Corn, soybeans, sorghum and tobacco are the most widely grown crops; hay, small grains and pastures are grown to a lesser extent. Only one farmer grows vegetables.

- **Cover crops/green manures used:** One farmer didn't answer, and three said they don't use green manures or cover crops. Overall, small grains are the most popular ones (grown by 11 farmers), followed closely by clovers (grown by nine farmers). Four of the farmers use alfalfa as a green manure/cover crop, three use hairy vetch, and one or two use rye, fescue, or an unspecified legume.
- **Crop that did best during first year chemicals were reduced:** Twelve farmers didn't say; three said hay, two said corn, and one each said alfalfa and small grains.
- **What happened to yields when chemicals were reduced/eliminated?:** Again, nine farmers didn't answer. Of

> *All three farmers who cited a reason for yield declines said weeds were the culprit. None said nutrient deficiencies reduced crop production.*

the nine who did, only one said yields declined, while the rest said they stayed the same.

- **Why did yields decline?:** There were only three usable answers, but, significantly, all three said weeds were a yield-limiting factor. One farmer also cited a lack of know-how, and another mentioned insects.
- **Which plant nutrients were a problem?:** Of the six farmers who reported nutrient deficiencies, the answers were fairly equally divided between nitrogen (cited by three farmers), phosphorus (two) and potassium (two). None mentioned lime or micronutrients.
- **What happened to net income?:** Nine farmers answered. Five said income remained unchanged, three said it increased, and one said it declined.

Southeast

Responses came from 16 farmers in Maryland, Virginia, North Carolina, South Carolina and Florida. Seven farm from 50 to 200 tillable acres, seven cultivate more than 200 acres, and two farm fewer than 50 acres. Corn, soybeans, forages and small grains again are grown on most of the farms; only three farmers raise fruits and/or vegetables.

> *Small grains are the most popular green manure/cover crop. They were also cited as the easiest crop to grow without chemicals.*

- **Cover crops/green manures used:** Of the 13 farmers who said they use green manures or cover crops, nine said they rely on small grains, six use clovers, three use hairy vetch, and three use alfalfa. Brassicas, fescue, rye and buckwheat are each raised on at least one farm.
- **Crop that did best during first year chemicals were reduced:** Hay and small grains again were mentioned as the easiest crops to grow without chemicals, though a few farmers had success with soybeans, tobacco and corn.
- **What happened to yields when chemicals were reduced/eliminated?:** Of the 13 farmers who answered this question, eight said yields of at least some crops stayed the same, while five said they declined. Only one farmer reported yield increases.
- **Why did yields decline?:** Only seven farmers offered reasons for yield declines. Weeds were cited by four, and a lack of proper equipment by two. One mentioned nutrient deficiencies.
- **Which plant nutrients were a problem?:** Four farmers didn't say, but the majority (nine) said nitrogen was their biggest nutrient worry. Only three said K was a problem, and one cited P.
- **What happened to net income?:** All but one said it stayed the same or increased.

Northern Plains

Responses came from 14 farmers in South Dakota, North Dakota and Montana. More than half farm at least 450 acres. The four largest farms exceed 900 tillable acres, and the smaller ones average about 200, with only one being smaller than 100 acres. Every one of these farmers grows small grains (usually wheat or oats), while eight grow corn and eight grow some type of forage legume (clover or alfalfa).

> *All but one farmer rely on green manures and cover crops to supply a portion of small grain crop nutrient needs.*

- **Cover crops/green manures used:** All but one of these farmers use green manures and/or cover crops: Nine use clovers, six use alfalfa, three use small grains, and one or two also raise buckwheat or rye.
- **Crop that did best during first year chemicals were reduced:** Not surprisingly, most farmers said small grains were the best crop to grow without chemicals. One also mentioned hay. Two said all of their crops did poorly when chemicals were reduced or eliminated, but didn't say when yields rebounded.
- **What happened to yields when chemicals were reduced/eliminated?:** Ten farmers answered this question, and half said yields of at least some crops decreased. An equal number said yields remained unchanged, and only one said they increased.
- **Why did yields decline?:** Seven farmers gave reasons, most offering more than one. Weeds were cited by five farmers, nutrient deficiencies by three, and equipment by one. One farmer checked the "other" category, and one said Murphy's Law was in effect: Whatever could go wrong, did.
- **Which plant nutrients were a problem?:** One farmer mentioned phosphorus, six said nitrogen, and the rest didn't answer.
- **What happened to net income?:** Of the nine farmers who responded, only two said net income declined, while the rest said it stayed the same or increased.

Northwest

Responses came from nine farmers in Idaho, Washington and Oregon. Farm sizes vary greatly: Three are from 50 to 100 tillable acres, three are from 150 to 200 acres, two are about 440 acres, and one is 1,500 acres. The crop mix varies, too. The only real consistency is that none of the farmers grow corn, while six grow small grains (wheat, barley or oats). A few of the farmers raise specialty crops like beets, potatoes, peas or lentils.

No farmers grow corn, and none said yields dropped when they cut back on chemicals.

- **Cover crops/green manures used:** Of the eight farmers who grow them, five use alfalfa, four use small grains and three use clovers. Rye and hairy vetch are also used by at least one farmer.
- **Crop that did best during first year chemicals were reduced:** Four didn't say, but the rest all said hay was their most successful crop. One also said small grains did well.
- **What happened to yields when chemicals were reduced/eliminated?:** None of these farmers reported yield declines. Four said yields remained unchanged, and three said they increased. Three farmers did not answer.
- **Why did yields decline?:** Though none of the farmers reported yield declines, one did say that nutrient deficiencies seem to be a problem.
- **Which plant nutrients were a problem?:** More than one farmer answered this question, which probably means that nutrient deficiencies did occur, but didn't limit yields to any great extent. Of the five farmers who cited nutrient deficiencies, three mentioned nitrogen, two mentioned sulfur, and one said potassium.
- **What happened to net income?:** Most said net income stayed the same or increased when chemicals were withdrawn. The lone farmer who reported a lower net income is the same one who mentioned that nutrient deficiencies occurred. Yet he also said yields did not decline.

Southern Plains

Only seven farmers from Texas and Oklahoma have filled out questionnaires, so far. Most of the farms are smaller than 100 tillable acres, though one is about 250 acres and one is about 700. All but two of the farmers grow wheat. Barley, milo, grasses, peas and lentils also are grown on some of the farms.

- **Cover crops/green manures used:** Three of the seven farmers do not grow them. Of those who do, three use some type of legume, one uses small grains and one uses brassicas.
- **Crop that did best during first year chemicals were reduced:** Small grains, hay and corn were all mentioned at least once.
- **What happened to yields when chemicals were reduced/eliminated?:** Only four farmers answered. Two said yields were unchanged, two said they declined.
- **Why did yields decline?:** Only two farmers gave a reason; both cited nutrient deficiencies, while one cited weeds, too.
- **Which plant nutrients were a problem?:** Four said nitrogen, one said phosphorus and sulfur. Two farmers did not notice nutrient deficiencies.
- **What happened to net income?:** Four said it either stayed the same or increased, while only one said it decreased. Two farmers did not answer this question.

'Read! Research! Experiment!'

That's the advice offered by Edward Rach and Tonia Edwards, who breed Simmental beef cattle on 120 acres near Hillsboro, Ohio. Fact is, nearly a dozen other survey respondents from various parts of the country had the same recommendation for farmers interested in reducing their production costs.

"The best advice I have for buying inputs is *always* leave a test-strip," noted cattleman Helmut Klauer of Wells, Nev. "Many times the fertilizer has . . . no response."

"Keep your pencil, calculator, yellow pad and common sense handy," agreed Mary-Charlotte Shealy, who raises beef cattle and registered Appaloosas in Fair Grove, Mo. "Be cautious of sales pitches. The organic line has its share of fast-talkers."

Finally, Ron Rossman of Harlan, Iowa, wrote: "I believe the list of possibilities is endless that one can do to reduce input costs." On his 360-acre, mixed crop-livestock farm, Rossman began using a combination of synthetic fertilizers, cover crops and composted animal manure to maintain productivity when he cut back on chemicals in the early '80s. He estimates that ridge-planting and mechanical weed control save up to $15,000 a year in chemical costs, and feels that his solar farrowing-nursery unit offers additional savings. "It all takes time and effort," he concluded. "I wish I would have, or could have, started . . . 10 years ago instead of last year. I used herbicides and anhydrous ammonia for 10 years. I'm 34 years old now."

Nutrient deficiencies–not weeds–are the main reason for yield declines.

Join The New Farmers' Network— And Help Yourself

To practical, field-proven ideas for taking charge of your farm. Fill out the questionnaire below, and you'll join the *Farmers' Own Network for Extension*, a growing group of farmers who have reduced or quit using purchased fertilizers and pesticides, and are helping others do the same.

NAME _____ PHONE # _____

ADDRESS _____ CITY _____ STATE _____ ZIP _____

Please tear out and mail to: FONE, The New Farm, 222 Main St., Emmaus, Pa. 18049.

1. Acres farmed _____
2. What crops do you normally grow and what are your average yields?

   Crop _____ Acreage _____ Yield _____

   _____ _____ _____

   _____ _____ _____

   _____ _____ _____

   _____ _____ _____

3. If you use a crop rotation, please describe it.

Check the answer that best applies to your farm. Check more than one box, if needed.

4. My soil type(s) is/are: ☐ Sandy loam/loamy sand   ☐ Silt loam   ☐ Clay loam/clay
5. My soil phosphorus levels are: ☐ Low   ☐ Medium   ☐ High   ☐ Don't know
6. My soil potassium levels are: ☐ Low   ☐ Medium   ☐ High   ☐ Don't know
7. I test my soil: ☐ Yearly   ☐ Every 2 to 3 years   ☐ Every 4 or more years   ☐ Never
8. I currently use these soil conservation practices:
   ☐ Cover crops   ☐ Crop rotation   ☐ Strip cropping   ☐ Terracing   ☐ Contour cropping
   ☐ Conservation tillage   Other (please specify) _____
9. My primary tillage device is: ☐ Moldboard plow   ☐ Chisel plow   ☐ Offset disk

   Other (please specify) _____
10. To supply crop nutrients, I currently use: ☐ Synthetic fertilizers   ☐ Organic fertilizers   ☐ Green manures/cover crops
    ☐ Fresh animal manure   ☐ Composted animal manure   ☐ Sludges   ☐ Organic by-products
11. The green manures/cover crops I use are: ☐ Alfalfa   ☐ Clover   ☐ Vetches   ☐ Small grains   ☐ Brassicas

    Other (please specify) _____
    ☐ I don't use green manures/cover crops
12. The livestock I raise are: (please list number after each)

    Dairy cattle _____   Beef cattle _____   Hogs _____   Sheep _____   Poultry _____   Horses _____

    Other (please specify) _____
    ☐ I do not raise livestock.

13. I sell the following crops off the farm: ☐ Hay   ☐ Small grains   ☐ Straw   ☐ Grains   ☐ Fruits/vegetables
    Other (please specify) _____

14. I want to reduce/quit using chemicals for these reasons:
    ☐ Cut production costs   ☐ Receive market premiums   ☐ Environmental concern
    ☐ Personal/family health and safety   Other (please specify) _____

15. I've been farming with reduced inputs for: ☐ 2 years or less   ☐ 3 to 5 years   ☐ More than 5 years
    ☐ Haven't started yet   ☐ Never used fertilizers/pesticides

16. I've made the following changes during this period:
    ☐ Reduced/quit buying fertilizers   ☐ Reduced/quit buying insecticides   ☐ Reduced/quit buying herbicides
    ☐ Started a crop rotation   ☐ Started using green manures   ☐ Started using cover crops
    Other (please specify) _____

17. When I reduced/quit using purchased inputs, my crop yields: ☐ Increased   ☐ Decreased   ☐ Stayed the same
    If yields fell, how many years until they returned to normal? _____

18. My crop yields declined during the transition because of:
    ☐ Nutrient deficiencies   ☐ Weeds   ☐ Insects   ☐ Plant diseases   ☐ Equipment problems
    Other (please specify) _____
    ☐ Yields did not decline

19. Most deficient nutrient: ☐ Nitrogen   ☐ Phosphorus   ☐ Potassium   ☐ Other _____

20. How was deficiency recognized? ☐ Tissue test   ☐ Soil test   ☐ Observation
    ☐ Other _____

21. The soil amendments I used during the transition were: (Please list) _____
    ☐ I did not use soil amendments.

22. The crop that did best in the first year I reduced/quit using chemicals was: ☐ Corn   ☐ Hay   ☐ Small grain
    ☐ Soybeans   Other (please specify) _____
    Why? _____

23. While changing to low-input farming methods, my labor needs: ☐ Increased   ☐ Decreased   ☐ Stayed the same

24. Changes in my farm management meant I had to buy the following equipment: (Please list) _____
    _____
    ☐ No new equipment was needed.

25. Most of my information on farming with fewer inputs came from: ☐ My own ideas   ☐ Extension agent
    ☐ Consultants   ☐ Salesmen   ☐ Other farmers   ☐ *The New Farm*
    ☐ Other farm publications (please specify) _____

26. Since I began farming with fewer purchased inputs, my net income has: ☐ Not changed   ☐ Increased   ☐ Decreased

27. Would you like to receive the results of this survey? ☐ Yes   ☐ No

28. What personal experiences do you feel would help other farmers interested in reducing input costs?

They're Already Farming More Profitably

Meet nine farmers who took charge of their farms years ago, and learn how their ideas can work for you.

THEY SAVED $60,000 ON P AND K

GOOD YIELDS FOR 10% LESS

HE CUT HIS CHEMICAL BILL IN HALF

FORGET FOXTAIL

HE *NETS* $60,000 A YEAR—WITHOUT BUYING FERTILIZER

THIS CORN YIELD CHAMP DOESN'T USE HERBICIDES

WHY GROW CORN?

BUILD A NITROGEN-PLANTING WEED KILLER

HE PLOWS DOWN 100,000 POUNDS OF *HOMEGROWN* N

REVERSING EROSION

18 YEARS OF TOP YIELDS—WITHOUT P AND K

WHO NEEDS HERBICIDES?

GROW LESS AND MAKE MORE

☐ **THEY SAVED $60,000 ON P AND K**/*A critical eye on soil tests helps these Maryland farmers keep fertilizer costs under control.*

MILLINGTON, Md.—In 1981, Dorsey "Jack" Owings discovered something on his land that increased his profits by $60,000. Was it gold? Hardly. Oil? Wrong again.

What Owings discovered was something he and his father, Dorsey Sr., actually had known for several years: Their 3,500 acres on Maryland's Eastern Shore is rich in phosphorus and potash from years of maintenance fertilizer applications. It's so rich, in fact, that in 1981, they were able to save $60,000 by eliminating potash fertilizer and reducing phosphorus applications by two-thirds, without sacrificing yields or crop quality.

Owings and Son Inc. is one of five or six huge, diversified agri-corporations that dominate the gently rolling landscape of Kent County, Md. Cash crops include spinach, wheat, tomatoes and sweet corn. Field corn and soybeans are grown to feed the farm's 500-head, farrow-to-finish hog enterprise. And, with two partners, the father and son operate a spinach packing plant that processes 200,000 bushels a year.

Such diversification has stabilized their income amid today's topsy turvy commodity prices, but it hasn't shielded them from soaring input costs that gobble up cash faster than you can say "foreclosure."

"Four or five years ago, I was paying $80 a ton for muriate of potash and $200 a ton for diammonium phosphate," says Jack, who does most of the actual farming now. "So even though our soil tests consistently showed P and K in the medium to high range, we followed the labs' recommendations and applied 80 to 120 pounds of each." Why? He shrugs, then forces out an answer: "Habit. That's the only reason I can think of."

But when fertilizer prices started rising—muriate soared to $150 a ton; and DAP to about $240 a ton—the farmers began seeking ways to break the fertilizer habit.

In the fall of 1981, a small article in a local farm tabloid got them started. "The article was by Dr. William Liebhardt," says Jack. "It was about some research he had done at University of Delaware. Basically, it said farmers might not be getting their money's worth out of fertilizer—especially when they applied it to *maintain* soil nutrient levels."

The farmers contacted Liebhardt, who is now assistant director of the Rodale Research Center in Maxatawny, Pa., and asked him to visit their farm.

"He came down, looked at our soil test reports, and told us we could get by *without applying any phosphorus or potash whatsoever*," says Jack.

As his father explains it, "We had a big discussion. Dr. Liebhardt told me to write to the Extension Service and ask them if they had any reports showing that maintenance fertilizer applications increased yields. I wrote, and they sent back lots of information, but none of it showed any yield increases for the maintenance applications the labs were recommending."

Jack pulls the letter from his files. Its opening sentence confirms Liebhardt's appraisal: "From our experience with test plots on Maryland farms, it is doubtful that you could expect a significant corn grain yield response from (fertilizer applications) when soil test is in the medium to high range." Getting vegetative growth off to a slightly faster start would be the only benefit of such applications, the letter adds.

Jack rummages through his file cabinet and produces another manila folder, this one containing soil test reports and fertilizer recommendations for a 120-acre irrigated field that had been planted to field corn, sweet corn and spinach over the last few years.

"Here's a soil test from January 1980, for sweet corn to be planted in spring 1980," he says. "Agrico's Agronomic Services Lab (Washington Courthouse, Ohio) lists both phosphorus and potash levels as *good*, yet they recommend adding 70 pounds of phosphorus and 120 pounds of potash for an expected yield of five-and-a-half tons of sweet corn.

"I applied 175 pounds of phosphorus and 250 pounds of potash, because that's the least my planter and spreader truck will handle."

Jack reached his yield goal, then tested the field in the fall of 1980, for a spinach crop to follow.

"P and K levels were still *good*," he says, "but this time they recommended *140 pounds of phosphorus* and 40 pounds of potash, which I applied in mixed fertilizer. I got a good spinach crop that year—about 3 tons."

Using the same soil test, he planted field corn the next spring, applying 175 pounds of phosphorus, but no potash. "I skipped the potash just for convenience," he says. "They had only recommended 40 pounds the year before." Late planting reduced corn yields to 120 bushels, he says.

"In fall, 1981, I let Agway take soil samples for that field. They sent them to A&L Eastern Labs in Richmond, Va., who recommended 210 pounds of phosphorus and 180 pounds of potash. That's when I saw the article and contacted Dr. Liebhardt."

When Liebhardt told him not to apply any phosphorus or potash, Jack was understandably skeptical. "Sure I was edgy," he recalls. "I tried discussing it with Agway, but I couldn't get them to agree with Dr. Liebhardt's recommendations." Why not? "Well," he says hesitantly, "I know, but I'd rather not say."

Tissue Tests Provide 'Spot Check'

After talking with Liebhardt and his Extension agent, Jack decided to apply no potash and just 30 pounds of starter phosphate fertilizer to his 1981 sweet corn.

"Yields fell to four-and-a-half tons," he says, "but I don't think the low fertilizer applications had anything to do with it. Three inches of rain fell shortly after planting, and that washed away my herbicide."

Eager to prove his theory, and to find out just how much P and K the corn crop actually had removed from the soil, Jack ordered tissue tests from Agronomic Services Lab.

"I'd never done tissue tests before," he says, "but they didn't question me. Hell, they'll do anything to sell you fertilizer."

Both phosphorus and potash levels were in the *high* range.

"That proves there was no shortage of nutrients," he says. "I'll probably continue to do tissue tests from now on. They give me a good spot check on the nutrients available to the following crop."

In 1980, again following lab recommendations, Jack applied 80 pounds of phosphorus and 160 pounds of potash to 180 acres of irrigated field corn.

"I got 140 bushels," he says. "Then, the following year, I applied 80 pounds of phosphorus and *no potash* to the same field, and got 170 bushels.

"This year," he continues. "I put on just 25 pounds of phosphorus and *still no potash*, and got five-and-a-half

100,000-bushel grain complex stores corn and soybeans for Owings' hogs.

tons of sweet corn—my normal yield. And the tissue tests were nearly identical to the other field."

He even divided one 240-acre dryland cornfield into three test plots. To one, he applied 250 pounds of phosphorus; to another, 25 pounds; and to the third, no phosphorus at all. None of the plots received potash.

"There was very little difference in yield between the three plots," Jack says. "In fact, the plot that got no fertilizer whatsoever yielded 105 bushels—highest of all."

Shortly after the 1982 corn harvest, Jack met some neighboring farmers at a local no-till conference. "I told them what I'd done, and they were very interested," he says. "But nobody's knocking down my door. I think the fertilizer companies have these guys baffled. They figure they've done alright for years, so why change now.

"Don't get me wrong," he adds quickly, "I don't think the labs and fertilizer dealers are out to cheat anybody. But they have to cover themselves. If they say 'Don't put on any fertilizer,' and yields are poor, who's the farmer going to blame?"

Jack cautions farmers not to reduce fertilizer applications arbitrarily. "You've got to have a history—at least a couple of years—of soil tests showing consistently high nutrient levels," he says. "Our soil samples are at the lab right now. When the results come back is when I'll decide what to do about P and K. But I'll tell you this, if they test out in the medium or good range, we won't be putting any on."

Adds his father, "You know, there are all kinds of farmers out there. It took us a while to accept the idea (of not fertilizing for maintenance). The corn didn't look good at first, but when it came time to harvest, we couldn't tell a bit of difference."

Of course, when it came time to pay the bills, they noticed a big difference.

Jack Owings outside one of two hog-finishing buildings.

☐ GOOD YIELDS FOR 10% LESS/
That's the bottom line on the Brubaker farm, where the management rule is self-reliance.

KUTZTOWN, Pa.—Trying to grow more corn on five acres than anyone else in Berks County is the type of farming Ben and John Brubaker have no use for. They know that aside from massaging the egos of the winners, about all Five-Acre Corn Club competition does is run up fertilizer, herbicide and insecticide bills, usually at the expense of other parts of the farm.

It's a good thing for the regular yield competitors that the Brubakers feel the way they do. For example, if the best cornfield on the Brubaker farm had been entered in the 1978 county yield contest, it would have placed second, only a few bushels behind the winner.

On top of that, the Brubaker corn is a lot more profitable than ordinary contest corn. Few, if any, herbicides, synthetic fertilizers or insecticides are used to produce it. Virtually all nutrients came from poultry and steer manure and alfalfa that grew in fields the year before.

By minimizing the use of all purchased inputs, the Brubakers keep cash operating costs up to 27 percent less than average for some crops. Overall, the bottom line of their beef, hog and chicken farm is a 10-percent reduction in production costs.

The Brubakers consistently produce yields that are as good as—often better than—county and state averages. And that's in one of the five top-producing counties in Pennsylvania.

The Brubaker farm is adjacent to the 305-acre Rodale Research Center (RRC). It consists of 72 acres Ben Brubaker bought and began farming in 1949, some 45 acres rented from a neighbor and 220 acres rented from the research center.

"Nitrogen requirements for corn production can be supplied almost entirely by on-farm sources without imports," concludes Victor A. Wegrzyn. As part of his doctoral thesis at Pennsylvania State University, Wegrzyn tested crop response to added nitrogen fertilizer (ammonium nitrate, leather dust and poultry manure) on the Brubaker farm in 1978, 1979 and 1980. He applied up to 250 pounds of nitrogen per acre on test plots.

"In general, supplemental N applications resulted in luxury consumption, rather than increased yields," Wegrzyn reports. "Increasing plant population increased yields as much or more than supplemental N. Soil N-supplying capabilities ranged from 99 to 221 pounds per acre and averaged 148, 147 and 132 pounds per acre for first-, second- and third-year corn after alfalfa, respectively.

"Overall productivity of this farm is high and with refinement of the management system could be increased," he adds.

The Brubakers' other accomplishments, according to scientists, include:

• Perfecting a rotation that provides a potential cash crop from every field, every year. The farm also produced

276 head of beef, 55 hogs, 5,235 dozen eggs and about 89 tons of hay for sale in 1982.

• Effectively controlling weeds and plant disease, almost entirely without chemicals.

"Weed control is excellent across the entire 138 hectares," according to Wegrzyn. "The fields where no herbicides are applied on the Brubaker farm are visibly indistinguishable throughout most of the growing season from fields where herbicide is applied. Except for the fall appearance of spindly, thinly populated late summer weeds on non-treated soils, weed control is excellent across almost all corn fields. Very little, if any, yield loss would be attributable to weed problems."

In a typical year, RRC agronomists say, the Brubakers will spend less than $600 on chemical pesticides, mostly herbicides for corn and soybeans on non-Rodale land, and insecticides for use around their barns.

• Cutting soil erosion in half, while maintaining profitable yield levels on droughty, erosion-prone soils. The Brubaker farm is located in a shallow valley at the beginning of the foothills of the Blue Mountains. Although some of the cropland is fairly level, most is located on shaly hills with slopes as steep as 25 percent.

Soil loss on the Brubaker farm averages about 4.53 tons per acre, compared with the 8.89-ton loss expected on the typical farm in the area. Brubaker soil losses range from a low of .77 tons per acre on an 11-acre field to an "intolerable" high of 13.83 tons on one 8.8-acre tract.

"There are some adjacent farms that are having erosion rates that are much higher than the Brubakers are experiencing," says Duane Pysher of the Berks County Soil Conservation Service. The neighbors, he explains, have rented large tracts of land and raise cash grains, usually corn and soybeans in continuous monocultures. Diversified livestock producers like the Brubakers and area dairymen have stayed with more soil-conserving rotations.

"There are no miracle secrets in what the Brubakers are doing in any way," Pysher adds. "Many of the things that they are doing are standard in agronomic aspects, as well as soil conservation practices."

To fully appreciate the Brubakers' strong commitment to soil conservation, it's important to note that their 337 acres are divided into about 95 fields. The exact number of fields may vary from year to year, depending on contour strip-cropping patterns. Each field is about 3.5 acres.

"It would all be bigger if we had flatter land," explains 32-year-old John Brubaker, who is taking over the family farm after his father's semi-retirement last year. "Five to 10 acres would be nice. You wouldn't have to pack up your gear and head for the next field so soon.

"It's not that it's any handier," he says of the crazy quilt of small fields. "It's just that that's the only way we can farm this land and take proper care of it."

Stewardship is vital to the Brubakers. John's father, Ben, set up the rotation system in use today when he began farming his original 72 acres nearly 35 years ago. When Ben began farming Rodale land in 1973, he started building it up with poultry manure, his rotation and other conservation methods. "Previous tenants had depleted the soil fertility and allowed considerable erosion to continue with no conservation effort," according to Wegrzyn. "The first crops in 1973 produced embarrassingly poor

John Brubaker combines wheat. The farm's small grain yields are equal to—or better than—state and county averages, but cost a lot less to grow.

yields." But by 1978, the elder Brubaker was producing more than 120 bushels of corn per acre on the Rodale land, which was also honored by SCS that year as one of the five best examples of soil conservation in the county.

"Their time frame for stewardship of the soil's productivity is measured in generations, rather than years," Wegrzyn says of the Brubaker family. "They farm to preserve the family's independence and the stability and continuity of the church (Mennonite) community. They are highly conscientious with soil conservation.... Their business management is based on a long-term planning horizon that cultivates a reluctance to take on debt, except for major purchases and strives to keep out-of-the-pocket expenses to a minimum. They would rather substitute labor for capital and, therefore, are willing to work harder and longer hours than is typical of other farms in the area."

Unlike many farmers who seem to spend all their money on new and bigger machinery and buildings, Ben Brubaker used some of the profits from his farm to help his children get started in farming in Wisconsin and Ohio. To help keep costs to a minimum, he bought mostly used, older equipment, building up a wide assortment of machinery that today includes five tractors, a crawler and a self-propelled combine. Much of his machinery dates from the late '50s and early '60s. But Ben and John, who is a highly skilled mechanic and welder, were able to "fix, adapt, or, in some cases, manufacture the needed parts or equipment so that the farm continued operating relatively smoothly," according to Rodale researchers.

The Brubakers' rotation also helped minimize production losses due to equipment breakdowns by spreading out the work load more evenly from March through November. While the exact cropping sequence depends on field and soil conditions, weed, insect and disease problems, red clover hays are generally grown for two years and alfalfa for three to four years.

"When hay fields are questionable, we'll give 'em that chance for spring growth," explains John. "Then, if they don't do well, we will manure them and plow them down. Hay after hay doesn't work so well. We save the manure for the corn. It needs it more."

Corn is grown for two, sometimes three years for grain and silage. Each year, John says, there is less nitrogen in the soil and weeds become more established. But insects and disease are seldom a problem. "It seems like we have the insects pretty well plowed down with the rotation. And it's been a long time since we had any blighted corn," he adds.

Soybeans and small grains follow corn. Wheat, rye, spring barley/oats are used as nurse crops for newly seeded hay fields. About two-thirds of the hay is sold off the farm. Rotation and regular mowing control weeds in small grains and hay. Early weeds in corn and soybeans are controlled with one or two rotary hoeings. Rows are cultivated once or twice and a hill cultivator may be used for the final cultivation if weeds persist in the rows.

Table 1 / Most Brubaker yields equaled or surpassed state and county averages from 1978 to 1982.

Crop	Location	Average Yields
Alfalfa hay	Brubaker County State	3.33 T/A 3.0 2.8
Other hay	Brubaker County State	2.67 T/A 1.9 1.9
Barley	Brubaker County State	38.1 bu./A 52.8 50.0
Oats	Brubaker County State	57.8 bu./A 56.2 56.2
Rye[1]	Brubaker County State	29.0 bu./A NA 31.4
Wheat	Brubaker County State	36.6 bu./A 37.8 34.6
Corn-grain	Brubaker County State	108.0 bu./A 85.3 91.6
Corn-silage	Brubaker County State	14.3 T/A 14.3 14.9
Soybeans[1]	Brubaker County State	38.1 bu./A NA 30.2

[1]Berks County, Pa., does not issue rye and soybean yield figures. Instead, they are calculated on a state-wide basis by the Pennsylvania Crop Reporting Service.

Table 2 / The cost of purchased inputs on the Brubaker farm is kept low with homegrown seed, on-farm machinery repairs and minimal chemical use.

Input	Unit	Range	Mean
Seed[1]			
Corn	bu.	16–24	18.8(4)[2]
Soybean	bu.	0–40	10.0(4)
Alfalfa	bu.	0–5	3.0(4)
Oats	bu.	10–50	20.0(4)
Barley	bu.	30–50	40.0(2)
Wheat	bu.	0–60	20.0(3)
Fertilizers			
Liquid Starter[3]	gals.	283–1,009	599
Chicken Manure	tons	144–216	181
Chemicals[4]	$	354–1,029	565
Gasoline	gals.	1,820–2,492	2,214
Diesel Fuel	gals.	2,416–3,205	2,723
Oil	gals.	55–223	152
Repairs (Parts only)	$	3,667–5,823	4,631

[1]During some years, the farmer used his own seed.
[2]Numbers in parentheses indicate number of years of data available for mean.
[3]Liquid starter fertilizer (9–18–9 analysis) was used on non-Rodale land.
[4]Chemical costs are primarily for herbicides for corn and soybeans on non-Rodale land, but also for barn insect sprays.

Agricultural experts may argue forever about how much money the Brubakers actually save by using older farm machinery and doing their own repairs. But no matter how you figure it, the savings will be substantial.

Brubaker Equipment	1982 Prices New	1982 Prices Used	Total Annual Ownership Cost[1] New	Total Annual Ownership Cost[1] Used	Estimated Annual Use/Hours	Ownership Costs/Hour New	Ownership Costs/Hour Used
Tractors							
105 HP Diesel—1850 Oliver	$35,200	$7,000	$5,632	$1,120	250	$22.53	$4.48
70 HP Diesel—1650 Oliver	21,900	3,500	3,504	560	400	8.76	1.40
50 HP Crawler—International Harvester TD 6	15,800	3,400	2,528	544	75	33.71	7.25
50 HP Gasoline w/Loader 77 Oliver	17,300	4,250	2,768	680	300	9.23	2.27
50 HP Gasoline—770 Oliver	16,300	2,100	2,608	336	300	8.69	1.12
45 HP Gasoline—77 Oliver	14,100	250	2,256	40	75	30.08	0.53
2 Manure Spreaders—New Idea	10,800	2,800	1,728	448	175	9.87	2.56
Plow, 4-Bottom Rollover—John Deere	2,900	1,000	464	160	75	6.19	2.13
Disk, Allis Chalmers	4,500	600	720	96	50	14.40	1.92
Springtooth Harrow	1,100	250	176	40	40	4.40	1.00
Cultimulcher—Brillion	5,200	1,550	832	248	75	11.09	3.31
Corn Planter—Massey-Ferguson	5,300	1,000	848	160	30	28.27	5.33
Grain Drill—Massey-Ferguson 33	2,700	1,100	432	176	40	10.80	4.40
Cultivator, 4-Row—Massey-Ferguson	2,000	700	320	112	100	3.20	1.12
Rotary Hoe	2,000	150	320	24	25	12.80	0.96
Sprayer	2,200	600	352	96	15	23.47	6.40
Rotary Mower	4,800	400	768	64	50	15.36	1.28
Mower/Conditioner—New Idea	6,800	2,000	1,088	320	150	7.25	2.13
Hay Tedder—Kuhn	3,185	3,185	510	510	60	8.50	8.50
2 Hay Rakes—Massey-Ferguson/New Holland	6,000	1,300	960	208	60	16.00	3.47
Pickup Baler	9,500	2,700	1,520	432	130	11.69	3.32
4 Hay Wagons	9,600	1,200	1,536	192	450	3.41	0.43
Combine—Massey-Harris	46,500	8,000	7,440	1,280	75	99.20	17.07
Corn Picker—New Idea	9,700	2,500	1,552	400	75	20.69	5.33
Silage Chopper—New Holland	9,100	2,500	1,456	400	75	19.41	5.33
2 Silage Wagons	13,600	2,000	2,176	320	140	14.54	2.29
Silage Blower—Allis Chalmers	3,400	450	544	72	35	15.54	2.06
2 Gravity Grain Wagons	2,000	1,350	320	216	125	2.56	1.73
2 Augers	2,000	1,200	320	192	50	6.40	3.84
Conveyor w/Motor	1,000	260	160	42	250	0.64	0.17
Totals	286,485	59,295	45,838	9,488			

[1] Annual ownership cost as a percent of purchase price includes 10 percent depreciation, 5 percent interest, 0.5 percent insurance and 0.5 percent housing.

The full text of the five-year Rodale Research Center study of the Brubaker farm is being published as a technical report of the Regenerative Agriculture Association Library. Single copies may be obtained by sending $5.95 to the Regenerative Agriculture Association, Publishing Division, 222 Main St., Emmaus, Pa., 18049. Summaries of the report are available free of charge by sending a self-addressed, stamped envelope to the same address.

Table 3 / **Despite the virtual elimination of chemical fertilizers, phosphorus and potassium levels on the Brubaker farm are high—and rising.**

PHOSPHORUS—Pennsylvania State University says phosphorus levels of more than 100 pounds per acre are "high."

	1977	1978	1980	1981	1982
Soil Type	Lbs./A Phosphorus				
Berks Shaly silt loam	171	216	156	158	179
Fogelsville Silt loam	228	373	261	307	399

POTASSIUM—Anything more than 150 pounds of potassium per acre is considered adequate for good crop nutrition by Pennsylvania State University.

	1977	1978	1980	1981	1982
Soil Type	Lbs./A Potassium				
Berks Shaly silt loam	203	219	235	313	266
Fogelsville Silt loam	149	203	227	313	360

On non-Rodale land, weeds in corn and soybeans may be controlled with a low rate of herbicide, especially if wet weather delays cultivation. "We use the minimum rate because of the rotation . . . so we can follow it with small grains or whatever we want. It's so we're a little bit freer to take care of the Rodale land with cultivation. With cultivation, you have to be pretty timely. Every acre we don't have to worry about means we can take better care of the next acre," John says.

"It gets a little hectic in the peak of the season, when first hay cutting and cultivation come together," he adds. "The other guy can plant his corn and go on vacation."

But it was the corn itself that took a vacation in 1983. Like much of the rest of the country, Berks County was hit hard by a drought. It was one of 18 Pennsylvania counties where 30 percent to 80 percent of the corn and hay crops were lost, according to state agriculture officials.

The Brubakers' yields were down that year, too. What corn they had was being harvested as silage. They had to buy grain that winter. But there's a feeling around their farm that things weren't as bad as they might have been. It's like 1980, when their rotation and lower plant populations on shaly soil helped the Brubaker farm produce 25 more bushels of corn than the county average.

Such lessons are not being lost on other farmers in the area, believes SCS's Pysher. "Their king crop, corn, got hit pretty hard (in 1983)," he says. "We may see more hay in the future. This is particularly true on soils of shale origin."

Crop rotations in small fields help the Brubaker's cut erosion in half and keep costs low.

☐ HE CUT HIS CHEMICAL BILL IN HALF/*And saves a dime from every dollar earned to find new ways of cutting back even more.*

NOBLETON, Fla.—Being a seed grower in a county full of ranchers doesn't bother Darryl Townsend one bit. In fact, the 33-year-old bachelor says it gives him an advantage over his competitors. "When you're isolated from traditional farm areas like I am, you don't have to deal with other people's pest problems," he explains.

But that's only part of the reason Townsend is able to raise nearly half a million pounds of seed and gross $300,000 a year on his 1,250 acres in south-central Florida. Using a chisel plow exclusively and following a strict crop rotation have kept his sandy loam soil in tip-top shape and cut his chemical bill by more than 50 percent. And reducing his acreage has helped him minimize equipment costs and manage the farm with just one hired hand.

"I used to spend $70,000 a year on *fertilizer alone*," recalls Townsend, who incorporated the Townsend Seed Co. in the early '70s on nearly 7,000 acres. Although his $50,000- to $75,000-a-year chemical bill is still his largest expense, he's convinced his crop rotation—two years of ZIPPER CREAM peas (a variety of crowder pea) followed by one year of millet and five years of Argentine Bahiagrass—results in better yields and less dependence on synthetic fertilizers and pesticides than he's ever had. "My goal is to farm the best I can," Townsend says. "When you get right down to it, I am still dependent on outside purchased inputs. But no matter how much fertilizer you use, you couldn't do what I do with this rotation."

Herbicides Seldom Needed

Townsend plants peas in August with an antique Cole planter mounted on a John Deere frame. Tests from the University of Florida have shown that legumes in his area require added nitrogen, so he applies a small amount of ammonium nitrate to the crop.

It's during the two years of row-planted peas that Townsend finds his only need for herbicides: He sprays isolated clumps of Johnsongrass by hand. "In this area, you really don't need herbicides," he points out. "With this soil, you can get 5 inches of rain one day, and be out in the field cultivating the next."

Second-year peas are harvested in November, and Townsend plows down the residue in spring. He says the pea residue supplies *all* the necessary nutrients for the heavy-feeding millet crop that follows.

Millet is planted with a homemade, 25-year-old cultipacker/planter, following his chisel plow. The assembly is hauled by his 125-horsepower International Harvester 1066 tractor. Townsend uses two 60-horsepower tractors for lighter cultivating and planting.

A large wheel in front of the cultipacker/planter opens a crease in the soil for seed, which is covered and pressed by a smaller wheel in back. Townsend says the device ensures planting at uniform depth, which is critical for millet's proper germination and growth. "If you plant millet too deep, it just doesn't make any yield," he explains. "It seems stunted."

Years of chisel plowing have improved his soil's water-holding capacity so much that the millet will produce a crop with very little moisture, he adds. In spring '83, for example, just one rain was enough to produce 700 pounds of seed per acre. That's 300 pounds less than he gets in a typical year. "But I would have gotten zero if the soil hadn't been chiseled," remarks the lanky graduate of the University of Florida's school of agriculture.

"The chisel plow is remarkable how it handles water and soil erosion," he continues. "It improves the permeability of the soil, and it helps water movement, both up and down."

Yet few of his neighbors in Citrus County seem willing to give up their moldboard plows. "I'm the only one around here who uses a chisel," Townsend says. "In fact, I went to the local equipment dealer for parts one day, and he didn't even know what a chisel plow was!"

The millet is harvested in June, and followed immediately with Argentine Bahiagrass, again planted with the chisel plow and cultipacker/planter assembly. Bahia, as Townsend calls it, is a deep-rooted, warm-season perennial used primarily as a permanent pasture for beef cattle in the Southeast. It is relatively free of diseases and insect pests, and performs best on sandy soil with a pH between 5.5 and 6.5.

Short, stout rhizomes enable it to become established on dry, sandy soils with no purchased fertilizer. But, to ensure adequate yields over the five years he grows the crop, Townsend fertilizes with ammonium nitrate, superphosphate, and sulfate of potash.

Townsend says a major advantage of Bahiagrass is that it completely eliminates the need for herbicides. "It chokes out all weeds," he notes. "It's a very selfish crop. That's one of the reasons it's so good in a rotation. And not only does it starve out weeds, it also cleans out nematodes. It's the top crop I know of for adding stable organic matter to the soil."

Native to South America, Bahia was first introduced into Florida in 1913. It produces seed for three to five years, then dies out naturally.

In its final year of growth, the crop is combined in September with one of Townsend's two Massey-Ferguson 750s, and the asphalt-like residue is rotavated and chisel-plowed in midwinter. As the organic matter decomposes, Townsend says it turns from a "nutrient robber" into a "nutrient provider," and helps buffer the soil against the acidulated fertilizers he's forced to apply to the following pea crop.

At the heart of Townsend's operation is a fully automated seed-processing plant, located in Dade City, some 30 miles south of his farm. With the help of his father, a retired custom harvester, Townsend processes 200,000 pounds of ZIPPER CREAM peas, 120,000 pounds of millet and up to 150,000 pounds of Bahiagrass seed each year for sale exclusively to wholesale jobbers.

His newly installed conveyor system has cut loading time by nearly 75 percent, and improved seed quality

Chisel plow and cultipacker/planter that Townsend uses to plant millet and Argentine Bahiagrass. Large wheel opens small furrow; rear wheel covers and presses seed.

immeasurably. In the old system, dump trucks with side-access hoppers had to auger the seed into bins for screening. It was a rough, time-consuming process that Townsend says resulted in a lot of crushed seed. Today, trucks just dump the seed into a huge, concrete pit, where a conveyor belt is waiting to carry them gently to the screening and packing machines.

Townsend applies all chemicals himself, but emphasizes that it's not a job he relishes. In fact, he spends about 10 percent of his income each year researching new ways to farm without them. Why? "Mainly because I don't like applying them. I don't like washing my hands 40 times a day, smelling them. It's just not a pleasant task."

In his own experiments, Townsend recently found that Furadan, a granular, soil-incorporated nematicide, dramatically improved his yields. Yet he no longer uses the product. "I think the (toxicity) of that stuff is understated," he says. "I've handled a lot of different chemicals, but you try to apply Furadan and your eyes will be twitching in 10 minutes."

Furadan is one of several popular nematicide/insecticides known for their ability to cause nervous disorders. The products are usually sold in granular form to reduce the risk of exposure to applicators. "You can't help but break the granules when you're loading it," cautions Townsend. "I don't particularly care to handle that toxic a chemical."

Townsend uses malathion and Sevin to control armyworms in Bahiagrass and stink bugs in peas, but doesn't apply them habitually, like many of his neighbors. "I'm not necessarily opposed to chemical use, but I spray mainly when I see there's going to be a problem," he says. "There have been many years when it (spraying) was unnecessary."

In his efforts to reduce chemical use, Townsend says he seldom takes his problems to the Extension service. "I've tried, but if it's any kind of problem like wanting an insect identified, all they usually say is, 'It's probably spider mites. Spray 'em with malathion.' That doesn't answer my question."

Despite all his successes, Townsend faces one hardship common to small *and* large farmers in every state: High interest rates have forced him to rent nearly 80 percent of his cropland. "When farmers get together now, it's not talking about farm practices, it's talking about credit," he laments. "The main problem I've had is financial, in that it's getting to where it's impossible for a farmer to own his own land."

The trouble is nationwide, he says, and it stems from absentee owners buying up farmland for tax shelters, then leasing it to farmers who feel no responsibility for maintaining it properly. "It takes many years to regenerate worn-out land," he explains. "If it's rented, how can you justify going through all that trouble? If you rent the land, you know you may only have it for one year. So you're going to mine all you can from it. Land today is being mined just like coal, oil or anything else."

His solution: "The government needs to redefine what a farmer is, not by *how much* he earns from farming, but according to what *percent of his income* comes from farming. Then the government should give farmers a tax break that will help them own their own land."

Meanwhile, he says, the time has come to stop emphasizing higher yields at the expense of environmental and natural-resource stability. "Any time you get a bumper crop, that's not natural," Townsend opines. "It doesn't matter if it's an organic system or a chemical one. It's natural for birds and insects to consume part of a crop."

He uncrosses his legs and leans forward like an ancient scholar about to reveal his innermost thoughts for the first time. "You know, farmers shouldn't accept things the way they are. There's always a better way to do things. I've learned some of my most important lessons from my garden. After all, that's all a farmer is—a big gardener. Right?"

☐ FORGET FOXTAIL/*Ridge-planting can help keep this weed out of beans for good.*

BOONE, Iowa—For the past few years, foxtail has been virtually absent in our ridge-planted soybeans. Getting this kind of control without herbicides depends on several important factors, including properly timed cultivations and the use of a fast-growing soybean variety with a thick canopy cover and excellent standability at high populations.

But the process really begins by allowing foxtail on the ridges to germinate. And that means leaving the soil undisturbed until you are ready to plant.

Let's begin with the building of ridges during late June cultivation of last year's corn. This operation brings foxtail seeds to the upper surface of the soil, where they are exposed to oxygen needed for germination. Many of the plants begin growing, but are smothered or shaded by the established corn crop, leaving a soil surface that is reasonably free of foxtail seeds.

The ridge itself should be about 12 inches wide, and fairly flat to allow the planter's sweep to scrape off 2 inches of soil from the entire planting surface. With narrow, pointed ridges, the weed zone is too close to the row.

Other factors to keep in mind when you're forming ridges are:

- The operation takes place while a crop is growing, so try to move as little soil as possible.
- Ridge-building should be completed as early as possible to avoid disturbing the crop's roots.
- The cultivator sweep or shovel should be as far away from the growing plants as possible.

The weed seeds that don't germinate, including those that may have dropped onto the surface, should be allowed to germinate before you plant soybeans the following spring.

Foxtail growing on the raised beds or ridges appears to have an allelopathic effect on broadleaf weeds in the lower levels of soil. But even more important, the 22-inch sweep in front of our Buffalo-Till planter's seed tube removes the top 2 inches of soil from the ridge, and pushes the growing foxtail, along with dormant seeds, corn ears and shelled corn, between the rows where the cultivator will control them. In addition to removing this residue from the seedbed itself, the sweep also covers foxtail that is growing between the rows.

If rainy weather delays planting, foxtail between the rows may get too large to be covered adequately by the planter's sweep. In this case, cut the foxtail with a cultivator, but be sure to leave the foxtail on the ridge for the planter to dislodge.

No need to worry about foxtail in lower soil layers. Foxtail germinates only near the soil surface, where oxygen is plentiful. The Buffalo-Till planter does not stir the soil very deeply.

Weed-Free Ridges

Soybeans are planted in a groove about 1.5 inches deep. The undisturbed soil in the planting area is firm, but not compacted. We've noticed foxtail in one area of our soybean field where the stacker wheels compacted the ridges while we were picking up cornstalks in early spring. Cornstalks in the rest of the field were stacked while the ground was frozen and less likely to compact; no foxtail was present in these areas.

Soybean variety selection is crucial to the success of a ridge-planting program. Varieties that emerge quickly and

Stirring soil with field cultivator or disk before planting beans allows foxtail and other weeds to germinate.

HOW TO FARM ON RIDGES

And keep your crops in the warmest, driest and most weed-free part of the field.

1. Build ridges at last corn cultivation in late June the year before planting beans.

2. No tillage between corn harvest and bean planting.

3. Buffalo-Till planter sweeps throw weeds and weed seeds from ridge tops into middles.

4. Plant indeterminate, fast-growing soybean variety—12 plants per row foot—after weeds emerge.

5. Rotary hoe prior to soybean emergence.

6. Second and third rotary-hoeings will be at five- to seven-day intervals. After that, the beans are large enough to cultivate.

7. On first and second (if necessary) cultivation, set disk-hillers as close to plants as possible. Large cultivator sweeps should be wide enough to undercut all weeds.

8. Reshape ridges on last cultivation (second or third, if necessary) in late June.

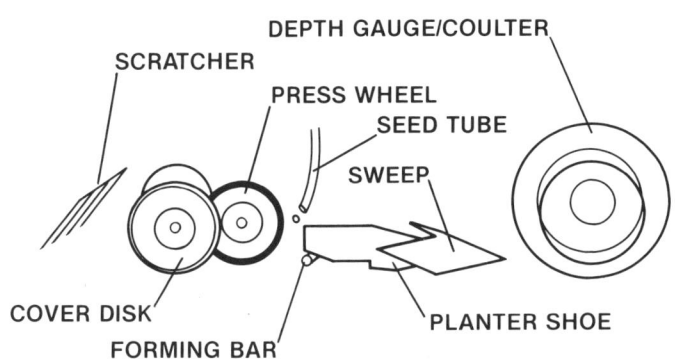

1. Planter Rig
Buffalo-Till planter rigged for planting soybeans.

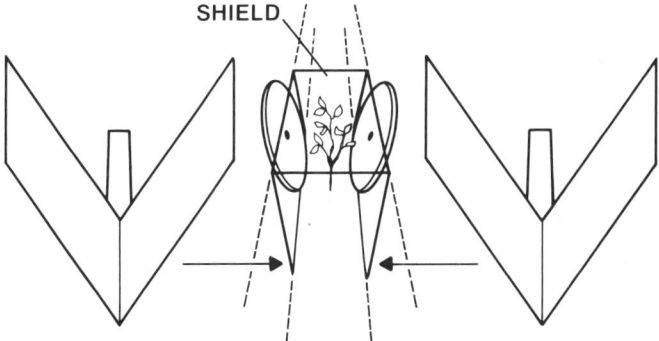

3. Cultivation
On 36-inch rows, 24-inch cultivator sweeps undercut weeds in row middles on first and second cultivation (if necessary). Disk hillers set 7 inches apart cut weeds close to beans, throw soil into row middles.

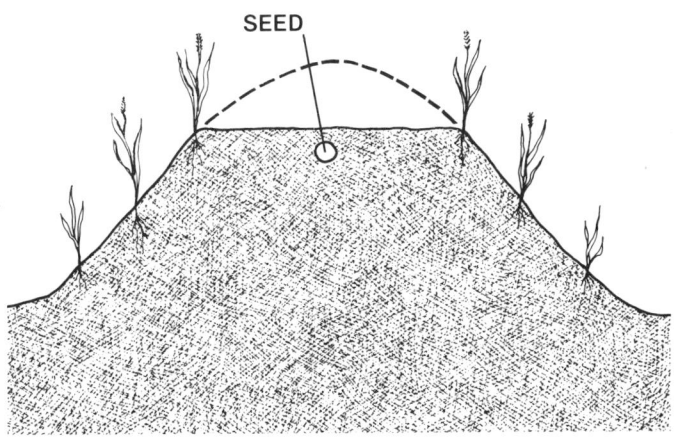

2. Planting
Smooth coulters at front of planter split ridges. Sweeps cut top 2 inches off ridges, moving weed seeds out of rows.

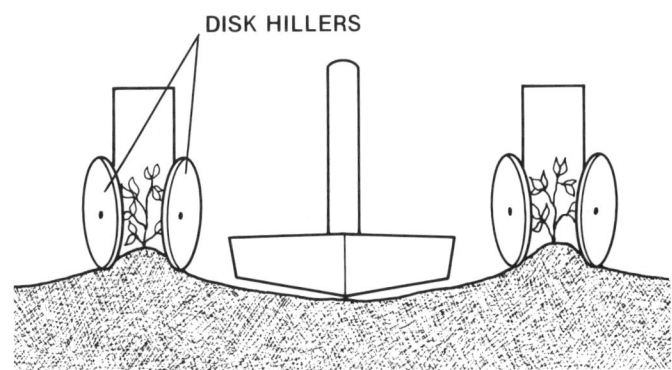

4. Rebuilding Ridges
On last cultivation, bedder shovels and disk hillers throw soil back into rows, rebuilding ridges. Shields protect beans from excess soil.

have a thick, full canopy shade weeds best. We tested four Group II maturity soybean varieties in 1983 for their ability to suppress weeds (see "What to Do When They Ban Herbicides," *The New Farm,* January 1984). All four—Asgrow 2680, DeKalb CX283, Prairie Brand 223, and Pioneer 2477—gave excellent foxtail control. We plan to test nine Group III varieties in 1984.

Podding characteristics are also important, since you'll probably follow beans with corn in the rotation. By choosing a soybean variety with plenty of pod clearance, you'll be able to form ridges for the corn during cultivations in June, without the danger of damaging soybeans. Ridges made after soybean harvest in fall tend to increase weed problems in the following year's corn.

Perhaps the most important characteristic to look for in soybeans to be used in ridge-planting is the ability to withstand high plant populations. Whether you plant in 30-inch rows or 36-inch rows, you'll need to sow 12 seeds per foot within the rows. This will produce between 175,000 and 200,000 plants per acre, populations that yielded the highest in tests by Pioneer Hi-Bred International.

Don't try to base your planting rate on pounds per acre, or you may end up wasting money or sacrificing weed control. Here's why: Let's say you're planting 12 seeds per row foot in 36-inch rows. That comes out to 174,240 seeds per acre, the recommended population.

If the seeds are large—say where 2,500 weigh 1 pound—you'd have to plant about 70 pounds per acre to achieve the desired population. But a smaller-seeded variety, in which 3,300 seeds weigh 1 pound, would only require 53 pounds per acre.

Planting the smaller seeds at the higher rate would produce 230,000 plants per acre (about 16 plants per row foot), which could easily cause the crop to lodge. What's more, the extra 17 pounds of seed would cost you $5.10 per acre.

Likewise, if you plant the larger seeds at the lower rate, you'll end up with less canopy cover and reduced weed control.

Generally, indeterminate varieties have a longer flowering period than determinate varieties, and yield more under a variety of environmental conditions. Soybean populations as high as 11 or 12 plants per row foot require a compact, indeterminate variety for optimum standability.

Finish Off Foxtail

Once soybeans are planted, the field should be rotary-hoed before either beans or foxtail emerge. The foxtail white roots that have germinated near the surface will be thrown out of the row to die.

The second and third rotary-hoeings will be at five- to seven-day intervals. After that, the beans will be large enough for cultivation. Like the Buffalo-Till planter, the rotary hoe doesn't stir the soil very deeply, so there's little danger of introducing oxygen to buried foxtail seeds. And, while the planter initially removes *most* of the weed seeds from the ridge, rotary-hoeing virtually guarantees a weed-free seedbed.

The high plant population, together with a fast-growing soybean variety that develops a full canopy, should shade any foxtail lucky enough to survive these tillage operations.

When the soybeans are established, use a heavy-duty cultivator with disk hillers and large sweeps to control weeds between the rows, and to rebuild ridges for the next year's corn crop. Begin cultivating when soybeans are tall enough so the rear sweep can push soil up against the row without covering seedlings.

Weeds between the rows should be cut out when they are 2 to 4 inches tall. *Before this, the weeds have been an asset;* they've been protecting the soil and taking up valuable nitrogen that was being released from the soil since early spring.

Disk hillers should be set as close to the soybeans as possible. The rear (24-inch) sweep has to penetrate deep enough to get under residue between the rows. One or two cultivations with this arrangement should control weeds between the rows.

Should any weeds remain, they'll be destroyed during the last cultivation when you rebuild the ridges. The bedder shovels are on the front bar of the cultivator, and the hillers are on the rear bar. These devices are set up to throw soil in alternating directions to cover as many weeds as possible. Bean shields protect the plants from excess soil thrown into the row.

The foxtail-free beans in the left portion of the photo above were ridge-planted into the cornfield below, in 1981. Both bean fields were cultivated two or three times.

Don Lambert disks in Austrian winter pea green manure crop that will supply his wheat with plenty of homegrown nitrogen.

☐ HE NETS $60,000 A YEAR— WITHOUT BUYING FERTILIZER/
This wise, old Washington farmer also gets good yields and cuts soil erosion by some 75 percent.

CHENEY, Wash.—While his neighbors spend up to $49.50 per acre every year just for nitrogen, wheat and dry pea farmer Don Lambert doesn't *pay* anything for his. He finds it just as easy and a whole lot cheaper to grow his own nitrogen.

All commercial fertilizers are a waste of money, he believes. He points to his farm as living proof of that. No synthetic fertilizers of any kind have been used on the 780-acre farm since the Lambert family began farming it three generations and 73 years ago. But the farm consistently produces 65 to 70 bushels of winter and spring wheat per acre—and nets a healthy $60,000 a year.

But Lambert's accomplishments don't stop there. A short, wiry man in his early 70s, Lambert comes from a long line of dedicated soil stewards. Although located in the Palouse area of far southeastern Washington—one of the most erosion-prone parts of the country, with yearly topsoil losses sometimes hitting 200 tons per acre—parts of Lambert's farm have up to four times less rill erosion than neighboring fields on the other side of his fence, according to a recent Washington State University study.

"It's a challenge," Lambert says of his life-long fight against erosion. "Sometimes that ground would need a slate roof to keep from washing."

The Palouse did have a "roof" once, a living thatch of prairie bunchgrass. It not only protected, but enriched the soil for thousands of years. The natural fertility of the area began 15 million years ago when volcanic activity blanketed the Pacific Northwest with countless tons of ash. Floods and glaciers later scoured away much of the ash deposit, but the 3,000 square miles of steep hills in the Palouse plateau were spared, and left with a yards-deep mantle of topsoil.

Early settlers discovered in the 1880s that the fine black topsoil was, amazingly, just as good on the hilltops as it was in the draws and valleys. Though early farmers had to contend with an ornery weed they call "Jim Hill Mustard" and slopes as steep as 30 degrees, wheat yields ranged from 40 to 60 bushels per acre. The topsoil was thought to be inexhaustible. And farming the Palouse hills began in earnest.

Depleting Soil 'Savings'

But it took just 100 years to find the soil's limits. Bald spots of light-colored clay now crown some Palouse hills where the once-bountiful topsoil has been stripped away.

Don Lambert doesn't need such statistics to realize the seriousness of the problem. It touches his daily life. Silt-laden waters flow into Bonnie Lake near Lambert's farm. Fishing used to be good there, says Lambert, who caught trout, bass and even pike in the lake. Now, his favorite fishing hole is silted over and growing cattails.

Ultimately, responsibility for such devastating erosion lies with farmers, themselves, Lambert believes. That's why he is switching from winter wheat to spring wheat as his main crop. Winter wheat encourages erosion because most farmers precede planting with clean, fall tillage, which leaves the soil vulnerable to winter weather. An estimated 70 percent of the area's 20 to 30 inches of precipitation falls during the non-growing season.

"January is the worst time of year for erosion," Lambert says. "It'll decide to get warm for a couple of weeks, so you've got snow melting, rain falling, and the gullies begin to form.

"(In 1982), the ground was frozen a good 12 inches down. When the rain hit, we got four-and-a-half inches in two days. You should have seen the washing." Under such conditions, experts say soil loss may exceed 100 tons per acre.

Lambert's second line of defense against erosion is also the source of his remarkable independence from purchased fertilizers. His Austrian winter pea green manure crop not only helps control erosion, but feeds his crops. The peas capture atmospheric nitrogen, bring other nutrients to the soil's surface, and build soil organic matter, which in some of Lambert's fields is as high as 4 percent. They also help control weeds.

He plants Austrian winter peas in early April and lets them grow until late July or early August, when they have 3 feet of growth and are in late bloom. If spring wheat is to follow, Lambert plows on the contour just enough to kill the cover crop, leaving a heavy surface residue and little chance for winter erosion.

'No Nitrogen Worries'

"We had the Austrian winter pea ground tested one year, and it had enough nitrogen for 84-bushel wheat," Lambert says. "All I know is that I don't have to worry about nitrogen when I plant wheat following an Austrian winter pea crop."

He doesn't worry about potassium or phosphorus, either. Potassium is so abundant in Palouse soils that virtually no one applies it. As for phosphorus, researchers and Lambert, himself, aren't sure, but they suspect Austrian winter peas somehow unlock this nutrient from the soil, making it available to following crops.

"Soil organic matter, total N, extractable P and extractable K tend to be higher in the top 12 inches of soil from the organic farm (Lambert's) as compared to the conventional farm," according to a study of Lambert's farm by Andrea Weilgart Patten, a Washington State University graduate student.

Lambert grows all his Austrian winter pea seed. It takes 12 to 15 acres to produce enough seed to plant 150 acres to the green manure crop. His seed crop is planted in April,

The Lambert farm—good yield, little erosion and no purchased fertilizer for three generations.

Soil tests showed that Lambert's legume cover crop supplied enough N for 84-bushel wheat.

harvested in September. The exact time of harvest is not critical. "There's very little shattering. I just go in when I'm sure the seed is mature," Lambert says. He direct combines pea seed with an International Harvester 403 combine set up for legumes. Normal yield is 1,000 pounds of seed per acre.

Lambert's belief in self-sufficiency and frugality extends from his machine shed to his kitchen. He pulls his implements with two "old and reliable" Caterpillar tractors. He paid cash for both. He doesn't believe in credit. "The only time I ever used credit was to buy a new 1929 Model A Ford. I never knew how fast the end of the month could come! I swore I would never use credit again. And I haven't." The family also keeps chickens to supply eggs, and a few head of beef for meat. "Some of the farmers around here run to the grocery store for *everything*," Lambert grouses.

Besides wheat and Austrian winter peas, Lambert grows dry, edible peas. The Palouse is known as the dry pea and lentil capital of the United States, and for good reason. It produces virtually all the domestic supply of these crops and exports some, too.

Dry peas and lentils use little, if any, of the nitrogen they fix in the soil, leaving ample nitrogen for the wheat that follows. They also break weed and disease cycles that could become established in continuous wheat. Lambert grows an old, favorite dry pea variety, ALASKA, which yields as much as 1 ton per acre. But the peas are not competitive with weeds and Lambert treats the crop with a selective herbicide, the only chemical he uses.

Finally, to give his soil as much protection as possible, Lambert plows on the contour wherever he can and disks only light stubble. He chisel plows all he can.

His concern for the future of the Palouse is shared by many farmers and agricultural researchers. For now, they would simply be happy to slow the huge soil losses and restore some balance to the system. That may happen. Some say it *must* happen—and within the next decade—if the Palouse is to continue as a center of food production.

With the wisdom of Lambert's traditional ways, a growing awareness of the need for better soil conservation, and soil-saving tillage systems now in the works, the Palouse may once again wear a soil-regenerating mantle of protective, yet bountiful grasses.

☐ THIS CORN YIELD CHAMP DOESN'T USE HERBICIDES / *And for two years he hasn't sprayed insecticides, either.*

CAIRO, Ga.—Ordinarily, farmers who win corn yield championships aren't known for their commitment to regenerative agriculture. After all, what's so great about a highly leveraged 'get-big-or-get-outer' pulverizing his soil with a moldboard plow, sterilizing it with high-priced, deadly pesticides, and then tripling his fertilizer application just to boost production of an already over-produced crop?

Lee adjusts Birchflex cultivator with Danish tines, which helps him produce record corn yields without herbicides.

Well, Carl Lee is one corn yield champ who doesn't quite fit that mold. Lee, 45, won the national Class A dryland corn yield title in 1981 *with no herbicides whatsoever.* And in the 1982 contest, sponsored yearly by the National Corn Growers Association of Des Moines, Iowa, he placed third nationally and first in his state—this time *with no herbicides — or insecticides.*

How did he do it? For starters, he applied about 225 pounds of nitrogen per acre, considerably more than the 150 pounds recommended by his Extension service for 100-bushel corn. But yields of 211.51 bushels per acre (1981) and 212 bushels per acre (1982) aren't guaranteed just by pouring on the nitrogen. The crop must be protected from weeds, insects and diseases. And most years, Lee is able to do that with no synthetic pesticides.

"If I can grow this kind of corn without herbicides, why spend my money on them?" asks the tall, shy Georgian. "That's part of what got farmers in trouble in the first place—spending money we didn't need to spend.

"Besides," he continues, "anything that will hinder the growth of grass is definitely going to have an effect on corn. I bought some 2,4-D a few years back, and it's still sitting out in the barn. Even with a little bit, you can notice some corn injury."

Avoiding herbicides may be one reason Lee's non-contest corn yields of 108 bushels per acre are nearly twice as high as the Grady County average for dryland corn. But he's quick to point out that relying on his chisel plow to keep his soil in good physical condition is equally important. He uses a moldboard only to work terraces on sloping fields. "The moldboard plow turns organic matter too deep and brings the clay soil to the top," he explains. "It gives you more erosion problems, because you have less organic matter in that turned-up clay. It's funny, though, the Extension people keep telling me it (depending on the chisel plow) won't work."

Soil Tests 'Come and Go'

Before he began farming in 1977, Lee worked at a Monsanto chemical plant in Cairo for seven years. "I saw a lot of soil tests come and go, and learned a lot of things some people don't know," he confesses, declining to elaborate further. As a result of the experience, he regularly applies more fertilizer—especially nitrogen—than necessary, preferring to build up soil nutrient levels for optimum yields. "In this area, we don't usually get a nitrogen carryover," he says. In the North, he adds, "the land freezes in winter and nutrients don't move as much. Our winter rains leach nitrogen right out." He also applies periodic doses of micronutrients.

It was Lee's 'nutrient maintenance program' that led a nearby Northrup King seed dealer to suggest he enter the yield contest. "They knew I was working on it and that something good was bound to happen," Lee says.

Participants in the contest are required to combine at least one and one-fourth acres of corn from a five-acre plot. The harvest must be witnessed by county agricultural officials, and the results are published in a winter issue of *Corn Grower* magazine.

Lee's first year in farming was less than successful. Two rainless months caused local corn yields to plummet to a depressing five bushels per acre. Lee lost $40,000 on just 250 acres. "There were more acres in this county that weren't combined than that were combined," he recalls. "We had a disaster."

Subsequent years were more successful, and Lee gradually built up his farm size to its present 350 acres, although heavy rains and his participation in the Payment-In-Kind program kept 180 acres idle in 1983. Further expansion, he says, is out of the question. "The average farmer is trying to get too big," he notes. "He overloads himself, and then has to spread his money too thin. One of the keys to any business is that you've got to keep expenses lower than your income."

Double-Duty Disking

Lee's non-chemical weed control program actually starts at planting time in February. He plants a fast-growing, 100-day corn variety on 30-inch rows. "You get a quicker canopy cover to help shade out weeds," he says. "And there's less water loss."

A Birch flex cultivator with Danish tines is used to control cocklebur, sicklepod, pigweed and wild morning glory. Lee says it works like a springtooth harrow with a large sweep in the middle of the row. He cultivates only once, in mid-March or early April when the corn is about a foot tall.

At the heart of his operation are two, well-timed diskings: one in July, immediately after harvest; and another in fall. "This way, the weeds haven't set seed yet and I can kill them before they make seed," he says of the first disking. Fall disking kills any late-germinating weeds, he adds.

Lee says most of his neighbors who use herbicides also cultivate twice. Others spray out of habit, even when weeds pose little threat to crops. "I can't see it. I can't afford it, either," he quips. "They end up making more trips over their fields with herbicides than I do without them. It just doesn't make any sense."

Lately, he's noticed more and more local farmers disking right behind their combines. "When you compare the cost of chemicals to the cost of disking, one extra disking is a lot cheaper than the cost per acre of applying an herbicide," he says.

When Lee won the national corn yield title in 1981, he applied a small amount of Furadan to control insects. "In this area, I think it would be impossible for us to get by without an insecticide," he says. "You (Northern farmers) don't have the soilborne insect problems we have." Still, Lee points out, his '82 and '83 crops did just fine without the potent pest killer, largely because stirring the soil right after harvest destroys the food and habitat of many pests. "I still think disking to kill all that leftover plant material plays a big role in insect and disease control," he says. "An insect that survives off live plant matter won't last long with dead plants.

"Plus, all that organic matter is good plant food and helps stabilize the soil. I've got no scientific figures to prove it, it just seems to be working."

He also uses pheromone traps to monitor populations of army worm, tobacco budworm and corn earworm.

The term 'regenerative farmer,' in its strict sense, may not accurately describe Carl Lee. In fact, he's still doubtful that this book's readers will find any value in a story about his admittedly chemical-intensive operation.

But in many ways, Lee represents a growing number of farmers, not only in America but throughout the world, who stubbornly defy the critics and achieve success by pursuing a goal of self-reliance and by limiting their acreage to manageable size. As Lee puts it: "Big isn't always better. Being a small farmer is about the only thing that's helped me to survive, lately."

"Let the farmer with the flat ground in Iowa or Illinois grow the corn," says dairyman Jack Kinsman. "These hills in Wisconsin were never meant for growing row crops."

☐ **WHY GROW CORN?**/ *It's a question dairyman Jack Kinsman feels more farmers should ask themselves. He quit corn in 1965, went to grasses and legumes, and is doing just fine.*

LIME RIDGE, Wis.—Jack Kinsman does not lead the life of an ordinary dairyman.

• Instead of putting in slavishly long hours tending cows and crops, Kinsman finds much time for off-the-farm concerns.

• Instead of worrying over dairy price supports, Kinsman's mind wrestles with social justice questions.

• Instead of having grabbed more acres and milking more cows over the years, Kinsman purposely tailored his farming operation to be small.

• Instead of nailing up no trespassing signs along his farm's boundaries, Kinsman encourages a steady stream of visitors to enjoy the beauty of rural life.

This farm, nestled in the picturesque hill country of southwestern Wisconsin, has become an inspiration for many who believe that small farms are not only beautiful, but bountiful, as well.

Kinsman and wife, Jean, have reared 10 children on a 150-acre farm, of which only 90 acres are tillable. They haven't had to depend on an off-the-farm job to make ends meet. Instead, combining a frugal lifestyle with excellent dairy husbandry practices, they have managed to do quite well.

An integral part of Kinsman's approach toward farming is not using chemical fertilizer or sprays. He hasn't used any in 19 years, ever since a nerve problem in his leg landed him in the hospital for a month. Suspecting agrichemicals were aggravating the problem, he quit using them, cold turkey.

Central to Kinsman's farming strategy is his belief that raising corn doesn't pay its way on a small dairy farm, especially in rolling ground country.

"Let the farmer with the flat ground in Iowa or Illinois grow the corn. These hills in Wisconsin were never meant for growing row crops and leaving the land wide open for erosion," Kinsman says.

"From what I can figure, it costs over $200 an acre for corn. Unless you have especially rich riverbottom ground, around here you will have trouble making a 100-bushel crop. I am better off to spend $2,000 to $2,500 per month to buy grain. In addition, I am protecting my land from erosion, and I do not have to go through all the headaches and worrying that go along with raising corn.

"The farmer small talk always centers around how their corn is growing. When they ask how my corn is doing, I tell them I do not grow corn. They just do not know what to say to that."

Kinsman raises grasses and legumes, instead.

"Hay is always in great demand around here," he says. "Every spring you'll find farmers who are desperate to buy hay. Good hay is practically impossible to find. Meanwhile, corn prices have been low for years, and I don't see any signs that this overproduction will be curbed."

Many Wisconsin dairymen complain they can't get a good seeding established, or that their fields winterkill after a year or two. The Kinsman farm lacks these problems. Kinsman's only hay concern in recent years has been to find enough mow space to store it.

Kinsman is a strong advocate of not growing pure alfalfa stands. "Nobody can convince me there's less nutritive value in those early-cut grasses than there is in alfalfa," he says.

The Wisconsin dairy farmer's seeding combination is 12 pounds of alfalfa to 1.5 pounds of orchard or brome grass to the acre. Half his fields are seeded to alfalfa/orchard, the other half to alfalfa/brome. Fields with orchard grass are cut first in the spring, then brome grass fields are ready to be cut.

The grasses do come back for second and third cuttings, although the alfalfa becomes more dominant later in the season. Kinsman feels the alfalfa/grass combination stops runoff much better than a pure stand of alfalfa.

Kinsman's hayfields look incredibly lush by late spring, with some of the grasses waist-high. Although not part of his seeding program, red and white clover come back year after year.

The dairyman pays his youngest children a nickel a plant to pull yellow rocket in the spring. They end up with small change. If they could pull across the fencelines, they would have half their college tuition money sitting in the bank.

Kinsman doesn't count every bale to boast of his yields. He does know, though, that he can depend on at least 8,000 60-pound bales every year. Plenty is leftover to help his son, David, who has his own farm, plus other farmer friends who run short on hay.

Kinsman's hayfields refuse to die out. Some have been in hay more than 14 years. They still look vigorous.

Kinsman credits his tillage program and barnyard manure for his field's longevity.

Back in 1965, when he quit chemicals and corn, Kinsman began a tillage program he calls "renovation." In August, Kinsman tills fields to be renovated with a chisel plow, which he calls a "digger." Chiseling in late summer, he believes, helps kill quackgrass, and does a good job of mixing organic matter into the top few inches of the soil. The process is done several times during the late summer. Fields undergoing renovation are also manured heavily. In spring, the fields are seeded with an oats cover crop.

Kinsman's hayfields are so strong that he went three years without getting the chisel plow out of the shed.

"This is my answer to no-till farming," he says.

Enough manure is generated by his dairy herd to cover all 90 tillable acres at least once during the year. The barn is cleaned every day during the winter, three times a week in the summer.

Kinsman buys no fertilizer. His barn lime (high-calcium lime from Michigan) supplies sufficient calcium, he believes. Also, long-rooted alfalfa pumps more up from the subsoil. (The nearby village was named "Lime" Ridge for a good reason.)

Kinsman takes pride in his meager investment in farm machinery. "Not growing corn has allowed me to get by with less than a $15,000 investment," he says.

Some of his farm machinery, like a trusty Massey 33 tractor that pulls the manure spreader, has been with him since he started farming 29 years ago.

Kinsman bales hay by hand, which is almost an anachronistic sight today. His baler, purchased for $250, works fine.

His Massey Ferguson 1080 tractor, bought for $8,000 five years ago, was his biggest equipment purchase. He also buys a new, 120-bushel manure spreader (New Holland) every three years because it's an implement he relies on almost daily.

What amazes Kinsman is how dairymen have gone silo crazy in the last two decades. He is especially opposed to the expensive, air-tight kind he derisively calls "those big, blue sauerkraut cans."

Kinsman built a 16- by 80-foot pit silo in 1963. It cost him $900. He feeds haylage or oatlage out of the silo from December to May.

"Some dairymen have forgotten that it's not tractors, barns or silos that make you money, but rather the cows inside your barn," Kinsman says.

The patient, steady work of milking cows twice a day is where Kinsman excels. Starting with a scrawny herd of cull cows on the rented farm he purchased, he has been able to build up a fine line of Holsteins—cows so handsome that they make farmers driving by in pickup trucks stop alongside the road for a second look.

Kinsman credits artificial insemination with being "one of the greatest things that ever happened to dairy farming."

On DHIA tests, his herd averages 16,000 pounds of milk and 600 pounds of butterfat. The Grade A milk is shipped off the farm daily.

Milking is done in a stanchion-type barn.

Kinsman believes a 35- to 40-cow herd is the optimum size for a one-man dairy. "Farmers my size can pay enough attention to their cows so that we can attain that 50-pound-per-cow average or better. Most of the big farmers I know have a tough time getting a 30-pound-per-cow average.

"The worst thing about big farming is that it leaves you unhappy. I'm on the Willow Creek Watershed Board. The big farmers who come to the meetings are worn out and gasping. They've got ulcers. It's terrible to see what shape they're in. Each one is trying to keep ahead of the other. Me, I'm 56 years old, and feel better than I did 20 years ago." Maybe not growing corn has something to do with that, too.

Kinsman believes baled hay is an excellent feed, because it supplies roughage for his cows. "All that soft feed from silos results in twisted stomachs and other weird cow problems," he maintains.

The Kinsman cows look especially clean because they are out in pasture from May until the snow flies. Fencing is done with electric wire, and pastures are changed two or three times a week. Keeping cattle contained in a small, muddy area only hurts herd health, Kinsman believes.

Kinsman's approach to farming is to study nature, then apply those principles to his operation. For example, his calves are fed only milk, rather than milk replacer. The low death loss for his young stock more than makes up for the milk that isn't shipped, he says.

Kinsman's 50-acre woodlot is his pride and joy. It has always supplied the heat for the family's home, as well as posts for fencing. Working closely with the county forester, Kinsman strives to maximize the woodlot's potential. Trees he planted on eroded wasteland when he moved onto the farm are now large enough to be logged.

Raising a family on a small dairy farm hasn't been an easy coast to prosperity. Yet it has more than fulfilled the Kinsmans' primary goal.

"Our intent was to make a living farming because we knew it was the best place to rear a family," Kinsman says.

In contrast to many other family farms, the Kinsman children have enjoyed the work routine that comes with milking cows. The two children who still live at home, Annette, 18 and Mark, 14, can handle milking chores by themselves.

Seven of the eight older children have gone on to attend college. They have introduced many of their friends to farm life by bringing them home for visits. Even now, with the older children embarked on their own work careers, they often come home to help on the farm, especially during the summer to make hay.

Kinsman is that unusual breed of full-time farmer who would rather show you his garden than his barn or machine shed. Three huge gardens, plus assorted fruit trees, provide much of the family's food.

Other food-related work projects, with everyone sharing the labor, have become fun outings. Pressing apple cider and sorghum-molasses have become part of the farm's tradition in the fall.

Up to 20 deer hunters gather at the Kinsman farm around Thanksgiving to hunt in the surrounding woods.

"Those of us who are fortunate enough to own land are only caretakers," Kinsman says. "We have an obligation to share access to the land with those who wish to appreciate its beauty and its values."

Kinsman has also become an ardent human rights activist over the years. A social awareness program called Project Self-Help and Awareness (PSA) had its genesis on the Kinsman farm. Now more than a decade old, PSA acts as an exchange program between whites in rural Wisconsin and blacks in Mississippi. Some 200 black youths come from Mississippi each summer to spend three weeks in the homes of Wisconsin residents. College-age Wisconsin students, in turn, spend time in Mississippi during Easter and Christmas breaks. More than 4,000 people have participated in this cultural exchange program.

The school buses that transport PSA participants are based at the Kinsman farm. Some challenging repair work, done primarily by Kinsman's oldest son, John, keep the old buses in shape for the 1,800-mile round trip to Mississippi.

Living on a small farm, appreciating nature, eating well to stay healthy, living right by showing concern for others, it all fits together in Kinsman's view.

"I wouldn't have peace of mind any other way," he says.

Annette Kinsman visits with the family horses.

Marilyn Kinsman helps her father with manure-spreading chores while home from her job at the University of Wisconsin in Madison.

☐ BUILD A NITROGEN-PLANTING WEED KILLER

MAXATAWNY, Pa.—Farmers around the 305-acre Rodale Research Center are used to seeing "different" equipment plonking around the place, but the latest creation to clatter out of our machine shop is a real traffic stopper.

A berm-side conversation might go something like this: "Now, just what do you suppose *that* is, they're driving now, Ben?"

"Dunno, looks like a grain drill on top of a cultivator!"

And that's what it is. The machine was developed to plant legume seed into a standing crop, such as corn. And why do that? For the more than 100 pounds of nitrogen per acre that legume growth can fix by the following spring, for one thing. The living legume sod also provides winter cover and erosion control, and adds soil-building organic matter (See "He Plows Down 100,000 Pounds of Homegrown N," *The New Farm,* July/August 1982). And with anhydrous ammonia prices expected to go nowhere but up, this ungainly cultivator/planter might become a familiar sight everywhere in a few years.

Every home-built machine is bound to be a little different, depending on the availability of used parts, but the following description can serve as a general construction guide.

The cultivator is a Pittsburgh two-row, three-point hitch model. One sold recently at a machinery auction near here for $80, but they can range in price from $50 to $350. The grain drill is an old, 8-foot "John Deere Van Brunt" with end-wheel drive. It cost $50.

Probably the most difficult part of joining the two machines was figuring out an alternative to the grain drill's end-wheel drive. Its wheels had to go because they would have taken up room in the row and taken downward pressure off the cultivator.

The solution was to mount a fluted coulter at the front of the cultivator frame. This coulter turns sprockets and chains, which lead to the grain drill drive shaft. "That coulter gets plenty of bite," says Bob Hofstetter, a research center agronomist. "It turned the grain drill drive just as surely as those big metal wheels would have." And an adjustable post holds the coulter at the correct soil depth, preventing skips.

To save the trouble of recalibrating the drill for every setting, the coulter drive system was geared to duplicate the turning speed of the wheels. The coulter first drives a seven-tooth sprocket, attached by chain to a large driven sprocket (15-tooth), which is attached by chain to small sprockets on the drill drive shaft.

It was easier to mount the grain drill to the cultivator frame. "I guess we were a little lucky, that drill was just the perfect size for the cultivator," Hofstetter says. The drill is held on by 10 bolts, which can be removed in about 10 minutes to free the cultivator. Cutting-torch and welding work made the match possible.

The grain drill is mounted above and behind the cultivator so it can meter seed down into soil prepared by the cultivator tines. Homemade frames extend down from the

This cultivator/planter "hybrid" seeds legumes or other cover crops into standing crops, such as corn.

drill to hold the seed hoses in place and to hold a set of drag chains. The chains help cover the seed and improve seed-soil contact.

Since the drill's original drive speed was kept, its original feed settings could be used. The drill has settings for flax, oats, barley, wheat, and field peas. Legumes used for overseeding are hairy vetch, red clover, and Austrian winter peas. "We set the drill for oats when planting vetch and it was right on the money," Hofstetter adds. The larger Austrian winter peas were metered out correctly on the field pea setting. Hairy vetch seed is about the size of a large BB.

Eight hoses carry seed from the gate to within five inches of the soil. Four hoses plant the full width of the middle, while two hoses on each "outside" end plant half of the outside rows.

For the best possible seed-soil contact, spring-mounted press wheels could be mounted behind the drill. This is especially true for the middle, which doesn't get wheel traffic to firm the soil on subsequent passes. Of course, with larger than two-row machines, even more rows would miss the firming action of the wheels.

The rig handles just like a cultivator, Hofstetter says. "Just travel at normal cultivating speed, you don't have to slow down. The seed will drop correctly, whether you're going one mile per hour or three." Ideally, overseeding in corn should be done when the crop is 6 to 8 inches high. But the exact time of the overseeding/lay-by cultivation will vary, depending on the weather and soil condition.

Wet weather in 1982 delayed one overseeding until corn was 30 inches high. Some root pruning and crop damage resulted. At the other extreme, dry weather might make the soil too hard for cultivation and overseeding. But most years, weed control needs, weather, and crop condition will permit an overseeding.

And for the effort, substantial amounts of nitrogen are fixed for the following crop, winter and early spring erosion are greatly reduced, and soil organic matter and moisture retention are increased, Hofstetter says. "Not bad for one cultivating/planting trip across the field and a few dollars worth of legume seed."

☐ HE PLOWS DOWN 100,000 POUNDS OF HOMEGROWN N/

Here's a nitrogen-fixing cover crop more and more farmers are using to beat rising fertilizer prices and soil erosion. They're also improving their soils, trapping moisture—and producing cash crops–the same year.

MILLSBORO, Del.—Bill Grantham gets 80 pounds of nitrogen per acre from his legume cover crop. And he grows the cover without taking any land out of cash crop production. He's using hairy vetch, a hardy winter annual that can be planted just after corn, soybean or vegetable harvest, or overseeded into standing cash crops. Grantham says it might not have made economic sense to grow vetch five or 10 years ago, but with chemical nitrogen prices nearing 30 cents per pound, "grow your own" nitrogen looks better every year.

"The vetch is just loaded with nitrogen when you turn it under," says Grantham, who is the farm manager for Townsend's Inc., a firm that includes an integrated broiler operation that processes 170,000 chickens each day, a feed mill, a soybean oil extraction plant and more than 7,000 acres in corn and soybeans. "This is our third year growing hairy vetch and it's worked well for us. I count on getting the equivalent of 80 pounds of chemical nitrogen per acre from it," he says.

Gary Clark of Clay City, Ill., got even more nitrogen from his vetch cover crop. He grew vetch on a 10-acre section of a 120-acre field. *He applied no chemical nitrogen to the vetch ground*, and 120 pounds of N to the rest of the field. "Come combine time, I couldn't tell a bit of difference," he says. He got 100 bushels of dryland corn per acre from the field.

Farmers and researchers across the country are proving that hairy vetch can fix plenty of nitrogen in the spring before planting time. And the vetch can grow in a wide range of climatic regions. "Hairy vetch is the most winter-hardy of the cultivated vetches and can be grown in most crop-producing areas of the U.S. It is adapted to light, sandy soils as well as heavier ones," according to "Forages," a standard reference on the subject. "In the northern U.S., only hairy vetch is sufficiently winter-hardy to be a winter annual; the other vetches must be sown early in the spring."

But increased demand for vetch—140,000 acres are now cropped to vetch in just Delaware—is driving up seed prices. In 1980, the seed cost only 35 or 40 cents a pound. It can go as high as 90 cents per pound. Even at this price, growing vetch can be cost-competitive with chemical nitrogen. "It'll cost me more than $20 an acre just to get the vetch established," says Clark, who finishes hogs and grows 500 acres of corn and soybeans. "All told, it'll be just a little cheaper than applying chemical nitrogen." But he's growing more acreage because the vetch has more benefits than just the added nitrogen, he says.

Harold Youngberg, Extension agronomist for Oregon State University and an expert on vetch seed production, agrees. "The (vetch seed) price increase is not necessarily a long-term trend," he says. As the price of seed goes up, the acreage in seed production naturally tends to go up, he explains.

And both Clark and Grantham are considering growing their own seed, which could be the ultimate solution to nitrogen self-sufficiency.

Bill Grantham shows off hairy vetch cover crop in early spring.

Some of those other benefits farmers and researchers attribute to hairy vetch are:

• The vetch cover crop acts as a nutrient pump, lifting P, K and other nutrients from lower soil levels and keeping them available near the soil surface.

• Vetch prevents soil erosion. Winter vetch cover isn't as thick as ryegrass, but it helps hold soil in place. Clark claims his soil loss is negligible.

• Whether it's treated as a green manure or a no-till cover, vetch increases soil organic matter.

• Vetch releases its nitrogen slowly and evenly to the cash crop, as opposed to a one- or two-shot chemical nitrogen application.

Corn yields on vetch cover-cropped ground are about 10 bushels higher than on fields fertilized with chemicals alone.

Just how does vetch cover-cropping work? Hairy vetch, *Vicia villosa*, is a hardy, fast-growing annual legume. In mild-climate regions, such as the Delaware-Maryland-Virginia peninsula, vetch can be planted just after corn, soybean or vegetable harvest and it will have enough time to become established and overwinter. In harsher climates, such as east-central Pennsylvania, the vetch must be *overseeded into the standing cash crop*, or it won't become established before winter.

"No one's been able to draw a strict geographical line," showing where a vetch cover crop can be established after corn harvest, says Jane Mountpleasant, a graduate student working with vetch at Cornell University in Ithaca, N.Y.

Grantham's high-clearance plow with 18-inch moldboards easily turns vetch under.

With either type of seeding, the vetch becomes established before cold weather sets in, then lies dormant through the winter. When warm weather returns in April, the vetch springs to life. "I'm almost afraid to walk in among those tendrils when the vetch really gets going," Grantham jokes. Then, it's just a matter of letting the rank vetch growth become knee-high and plowing it down for green manure or planting no-till.

In Delaware and south-central Illinois, vetch may be ready by the end of April or early May. Vetch may take until mid-May to reach the proper height in Pennsylvania or New York.

Grantham says he's been getting 80 pounds of N per acre pretty consistently from the vetch since he started growing it. He is very flexible with application of other types of nitrogen and he has many options. For example, he pivot-irrigates 1,930 acres of Townsend land with water that was first used in the broiler processing operation. That water contains some nitrogen and other nutrients. Manure from the poultry houses may also be applied to the fields. So some vetch cover fields may not need any chemical nitrogen, he says, while others may take from 30 to 60 pounds per acre to produce normal yields. He has tissue analyses of his corn done to determine nitrogen levels once or twice each season.

It's these kinds of figures that are causing the widespread adoption of vetch covers in Delaware, says Bill Mitchell, retired University of Delaware agronomist. More than 140,000 acres in Delaware are cover cropped to vetch now, he says. He has written Extension bulletins on the subject (available by writing Cooperative Extension Service, University of Delaware, Newark, Del. 19711) and has become something of a celebrity as the main proponent of vetch cover-cropping. He has been quoted on the subject of vetch in everything from the local newspaper farm pages to *The New York Times*.

Grantham says the University of Delaware is helping him and other farmers across the country by providing information on the basics of vetch cover-cropping, such as seeding.

Grantham plants vetch with a Vicon seeder, broadcasting 27 to 30 pounds per acre right after corn harvest. The combine is equipped with a stalk shredder so a mulch is left to help the vetch seed germinate.

"I could get by with a lower seeding rate, say 20 pounds per acre, if I used a drill to plant the vetch. But with the number of acres I'm planting, I don't have time to do anything but broadcast," Grantham says. Another option is to have the seed flown on. He says he gets better germination and good stands with his present system.

Clark, on the other hand, uses a drill and says he gets good stands seeding at about 20 pounds per acre after corn or soybean harvest. He uses all his vetch as a no-till cover.

Grantham decides how he wants to use the vetch cover in the spring. In 1982, he plowed down about one-fourth of the farm's 5,000 acres of vetch as a green manure.

No-till planting of corn or soybeans is done while the vetch is still alive. The vetch is then killed with an herbicide immediately after planting and remains on the soil as mulch. The herbicide costs $12 to $15 per acre and the total cost of no-tilling is a little higher than conventional tillage, Grantham says. But no-tilling saves time, and the mulch conserves soil moisture, so it's a trade-off, he adds. Grantham says he may not get as much total nitrogen from vetch with no-till as with plowdown, but with no-till, the nitrogen is released more gradually to the crop.

The vetch is no trouble to plow down, Grantham says. He uses a high-clearance plow with 18-inch moldboards to turn the vetch under.

One disadvantage of using a vetch cover crop is that the soil takes longer to warm up, says Ed Baker, who with his father and the rest of his family, grows corn, soybeans, potatoes and raises Holsteins and poultry in northern Delaware. In April '82 at the Baker farm, the soil temperature under vetch cover was 44 degrees F, while nearby bare soil was 50 degrees. He says vetch keeps the ground a little cooler than rye cover, which is also used extensively on the farm. But, so far, delaying planting hasn't been a problem, nor has it dented yields, he says.

Baker often plants vetch after potatoes because the potatoes come off early and are heavy users of nitrogen and organic matter.

Whether the cash crop is planted no-till or with conventional tillage, it benefits from the gradual release of nitrogen from the steadily decaying cover crop. Of course, the vetch also releases other macro- and micronutrients. These have been taken up by the vetch either from the soil itself, or from previous fertilizations, researchers say. The vetch serves as a reservoir for nutrients that might otherwise have been lost to leaching and washing.

Vetch can become a weed problem in small grains, however, even though the vetch cover crop is plowed down or killed before it can set seed, some seed that did not germinate at the original planting may volunteer years later. The severity of this problem seems to vary by region, says Bob Hofstetter, a member of the Rodale Research Center team working with vetch. Farmers who grow small grains may have to adjust their rotations to include more hay, corn or soybeans, he says.

Grantham says he has had little or no volunteer problems in barley planted after vetch.

Using vetch as a green manure crop isn't new. In 1940, Oregon's Willamette Valley had 84,000 acres in vetch seed production, says Don Brewer, Oregon State University agronomist. In 1950, the valley's 150,000 acres of vetch produced enough certified seed to meet a generous portion of the nation's nitrogen needs. The number of acres declined steadily through the 1950s as cheap chemical nitrogen displaced cover crops and legume rotations. Today, only 9,000 acres are in vetch seed production in Oregon, the leading seed producer. Oklahoma, Nebraska and California are also producing small amounts of vetch seed.

Now vetch is making a comeback as nitrogen prices increase. And this time northern farmers won't be left out of the legume boom because agronomists are developing new ways to establish vetch cover crops where the growing season is too short to plant vetch after corn harvest. And vetch can be planted in northern regions after shorter-growing-season crops.

Hofstetter says vetch can be flown, broadcast, or drilled (with a custom-made cultivator/planter) into standing corn. It germinates, grows right under the corn without reducing yields, then overwinters and is ready to be used as a green manure in the spring.

"We broadcast vetch following the final cultivation. So the time of overseeding depends on how much weed control we have to do," Hofstetter says. Both Hofstetter and Cornell researchers caution that vetch *interplanting* is still experimental and that it should be tried by farmers on that basis. Other legumes showing promise as interplanted cover crops are red clover, crimson clover and Austrian winter peas.

More farmers will turn to legume cover crops as a reliable, sustainable source of nitrogen as chemical nitrogen prices continue to rise, Hofstetter and others predict. And the new interplanting methods may enable farmers in any region of the country to grow a nitrogen-fixing cover crop and a cash crop in the same year.

Grantham also believes vetch will become more widely used. "It's not just for the nitrogen. You can go out to the field in the middle of summer thinking the soil is really going to be dry but you can push into the soil with your heel and then reach down and feel the dampness held there," Grantham says. "I'll be growing more vetch and trying it after soybeans."

Hairy vetch is lush and tall just before plowdown for planting.

Overseeded vetch already provides ground cover at harvest.

THEY'RE ALREADY FARMING MORE PROFITABLY

☐ **REVERSING EROSION**/*He just wanted to stop erosion. But this Iowa Farmer ended up eliminating pesticides, cutting costs, creating a steady cash flow—and building new topsoil.*

BELMOND, Iowa—Arden Kiefer's farm has no soil erosion! Matter of fact, Iowa State University agronomists say he will build one-half to one inch of new topsoil every five years. Kiefer is building his farm with hay. Not just a patch here and there, but his entire 180 acres of rolling land is one seemingly endless field of alfalfa. He's a cash crop hay farmer.

Kiefer says he converted his farm from row crops because he was just plain ashamed of the way he was treating the land after buying it in 1975. "I was ruining the farm!" he says. "In 1977, I was so depressed with the erosion I was getting. I terraced, but still with row crops I kept on destroying it. I made up my mind I was going to do something different. I thought about going with oats, but a friend of mine has oats, and he still has erosion. The only other thing I could think of was hay. I was going to stop the erosion, one way or the other."

Kiefer says some of the hills on his farm have 15-percent to 20-percent slopes. A 3-inch rain in just 12 hours would leave his hills scarred with gullies up to 10 inches deep and 8 inches wide. "The land all around here suffers the same thing," Kiefer says. Some farmers in the area do not even use contour plowing, he adds, they just plow straight up and down hills, and plant corn and soybeans. The Wright County Soil Conservation Service estimates the average soil loss with conventional row cropping in Kiefer's area is 10 tons per acre per year.

In addition to solving his erosion problem with hay, Kiefer also:
* Cut his machinery costs by 30 percent.
* Totally eliminated herbicides and insecticides.
* Broke into a new market that provides a steady cash flow.

I'm still in the learning stage," Kiefer says. "All I ever knew was corn and soybeans."

Right off the bat, Kiefer reduced his machinery investment from $100,000 to $70,000, replacing all his row crop equipment with two square-bale balers, one round baler, two self-propelled mowers/windrowers, one automatic round bale transporter, six 20- by 8-foot steel hay racks and seven small tractors. It's all new, except for the reconditioned tractors.

"If anything, I'm over-equipped," Kiefer says, "but I wanted the ability to get the hay up fast. I can get the 180 acres down and up in five days, if Mother Nature cooperates. And the extra machinery enables me to do it with up to 400 acres." He figures the average hay farm of 160 acres could be equipped for $40,000, about one-third of what would be needed for row cropping.

5-Year Crop

Establishing his crop put him ahead of the game, too. Seeding the entire farm to 87 percent alfalfa, 12 percent orchardgrass and 1 percent timothy, with oats as a nurse crop, cost $5,000. That sets Kiefer up for at least five years. After that, he figures he'll have to reseed. His maintenance cost is minimal, with potash fertilizer being his only expense.

"Every year I was spending $12,000 to $14,000 for seed, fertilizer, herbicides, etc. Now I spend around $3,000 for potash," Kiefer says. "That's half the cost of the fertilizer, alone, I was putting on for corn and beans. So really, as far as annual investment, it's much lower, and I have a cash flow most of the year."

He hasn't had to use insecticides or herbicides and, next to his erosion control, he couldn't be happier about that. "That's another thing that scared me. I've never been too much on herbicides, but in 1976, I went ahead and used them. I never saw anything like it in my life! Here you got this corn sticking out of the ground and there isn't a sign of anything else. Absolutely bare! That's scary to have something that powerful."

His land is clean. The alfalfa root system keeps weeds to a minimum. With mowing three times a year, a weed that does sneak through doesn't have a chance to reseed itself.

As with any farm operation, though, there are some costs involved in the business of hay. Kiefer says his two main expenses are labor and twine. "Right now, labor is high because I don't have an elevator system set up for the square bales," he explains. "It takes a lot of guys to relay them 14 high in tiers." He thinks he might use a little more fuel than a row cropper when all three balers are going, but he's not too sure about that. His biggest tractor is only 75 horsepower.

Obviously, round bales would be cheaper to package, but Kiefer believes small, square bales are the most marketable. As a cash crop farmer, his strategy is to fit the demand.

The actual baling runs smoothly with a crew of 12 to 14 men, many of whom are family; a son, a son-in-law and a whole bunch of nephews. Kiefer likes to cut 40 to 50 acres one day, cut a like number of acres the next day, then go back in the afternoon to bale the earlier cutting. And while the men are in the field, his wife, Sheryl, prepares meals, comparing the noon meal to the huge dinners made for threshing crews years ago.

Now that Kiefer has stopped erosion on his farm, his next priority is establishing a good "hay reputation" and marketing his product. After all, he can't expect the local elevator to buy his hay, like it did his corn and soybeans. There is no such thing as a hay marketing system in the Midwest, or anywhere else that he found. Minus any middlemen, he's had to set up firm ground between him and the farmer who buys hay. And he believes the basis for that firm standing is quality, fairness and consistency.

"It's going to take several years to build a solid rapport with farmers," Kiefer believes. "When Arden Kiefer says it's good hay, that reputation has to stand.

"What I want to establish is their confidence in the fact that I'll have the hay; I'll take care of my regular customers first, and that I'll be fair-priced," he says. "I don't care how

Stopping erosion is only one reason Kiefer no longer grows row crops. He's also cut his machinery costs by 30 percent, and eliminated pesticides completely.

high hay gets, my price will be stable during a given year, and the same to the guy in another state as it is to the guy across the road."

His advertising has been by word of mouth only. Kiefer figures his main market is within a 30-mile radius. He's not looking for big purchasers, but would rather sell his crop a few hundred bales at a time to individual farmers.

Although Kiefer pays strict attention to the quality of his hay, assuming that would be one of the first things customers would require, he's found it seems to make little difference, so far. After selling his first year's crop and much of 1982's, only one farmer was concerned enough to inquire about its food value. That customer was a dairyman. Regardless of the indifferent attitude, though, the best in hay is important to Kiefer, and he still believes it to be a requisite of many good farmers. To achieve that goal, he drilled the best seed he could buy, wanting a fine-stemmed plant that resists root rot.

He doesn't know the exact protein level of his hay, but says he does everything he can to assure high protein content. For example, he mows at mid-bud, even though top yield is at full bloom. He fertilizes twice a year knowing that every ton of hay has drawn 10 pounds potash from the soil.

After the first year's frenzy, Kiefer learned to take the rains in stride and be patient. In spite of things he read and what agronomists told him, he now knows that packaging hay at 24 percent moisture is not the thing to do. His new rule is 15 percent to 18 percent. Kiefer says he would rather have his hay rained on than put up wet by trying to beat the weather. All his hay, including round bales, is stored under roof in an L-shaped metal building that is 227 feet long, 60 feet wide, and 16 feet high. It holds 45,000 bales.

The first year's crop from just one cutting was light, only 4,000 square bales and 100 round bales. Kiefer says he could have sold that amount several times over. He isn't having to set up a roadside stand for his 1982 crop, either. His entire third cutting was sold right out of the field.

Increasing yield without sacrificing quality is Kiefer's next order of business. His four 1982 cuttings yielded a little more than six tons an acre. He figures six tons equals 100-bushel-an-acre corn selling at $3.50 a bushel. And with the price of corn predicted to remain unstable, Kiefer often does a lot better than anyone with 100-bushel or maybe even 150-bushel corn. Kiefer's growing a lot of good homegrown nitrogen, but vows he will never sink another moldboard plow into his soil.

He wants to get into livestock soon to complete the biological fertility cycle, and should have no trouble supplying his own stock with quality hay.

"We have just purchased another farm a half mile east of us," he explains. "I'm not going to be a bit bashful about putting that 160 acres in hay, too."

Now Iowa will probably never boast of any mountains, but if every speck of land could be farmed with erosion control like Kiefer's in mind, the state could again brag about knee-deep topsoil.

☐ 18 YEARS OF TOP YIELDS—WITHOUT P & K / *New research shows 'maintenance' applications of these nutrients are a waste of money.*

NEWARK, Del.—Preliminary results from an 18-year continuous corn fertilizer/yield study show that levels of phosphorus and potassium remained very high in ground that received *no* P and K fertilizer for nearly two decades.

Despite yields of up to 240 bushels per acre, there was no significant yield difference between plots that were heavily fertilized with P and K and plots that did not receive these fertilizers. The per-acre cost of unneeded P_2O_5 and K_2O at an application rate of 50 and 100 pounds, respectively, was $22.55 at 1984 prices. For a farm with 500 acres of corn, the total cost of unneeded P and K fertilizers for the 18-year period would have been $202,950.

"We should kick our old habits and take a fresh look at how we're using fertilizer and water," says William H. Mitchell, a retired University of Delaware agronomist. "When we began the study in 1965, it was intended to be a straight-forward look at maximum economic corn yields. We were in for a couple of surprises."

The experiment was conducted at the Delaware Agricultural Experiment Station, which is near Newark in the northermost part of the state. The farm's soil is a fine silt-loam. The field on which Mitchell's study took place was used for the production of legume-grass forage from 1949-1965. Mitchell says the field was not given any special treatment in those years. Soil tests taken at the beginning of the study showed high and very high levels of P and K.

In the spring of 1965, the entire 1.5-acre site was planted to the best field corn hybrid available. Some of the 37-by-48-foot plots were to receive no fertilizer of any kind for the duration of the study. Others received varying rates of nitrogen, but no P and K. Still other plots received one of four fertilization rates, ranging from 50 pounds nitrogen, 25 pounds phosphorus (P_2O_5), and 50 pounds potassium (K_2O), to 300, 150, and 300 pounds of NPK respectively. The plots were also split between irrigated and non-irrigated.

As expected, during the first few growing seasons there was no significant difference in yields between plots receiving nitrogen fertilizer but no P and K, and plots receiving the full NPK treatment. Mitchell recalls that yields for both were in the 150- to 160-bushel-per-acre range, irrigated, and about 80 bushels on the non-irrigated plots.

Yields Never Fell

But as more growing seasons passed, it became clear that yields on the plots receiving no P and K were not going to fall below those of their heavily fertilized counterparts. "The plots that weren't getting any P and K were hanging in there amazingly well," says Mitchell.

After a full 18 years, soil tests showed the plots receiving no added P and K still had very high levels of P and K. Those nutrients were still available for use by the corn plants, according to leaf tissue analysis and yields. In 1983, the final year of the study, the fully fertilized, irrigated plots yielded 230 to 240 bushels per acre, thanks to improved corn hybrids. *But the plots receiving no P and K also yielded 230 to 240 bushels.*

High Doses Useless

There was no significant difference in yield between plots that received even *maximum* fertilization—the equivalent of an accumulated 2,700 pounds of P per acre and 5,400 pounds of K per acre—and plots that received *no* P and K for 18 years.

Mitchell says he is not sure how and why the no-P-and-K-added plots continued to provide adequate nutrients for continuous corn. Much of the P and K must have carried over from earlier fertilization, but he theorizes that the soil itself may have yielded P and K, and that this may occur in other soil types.

The study helps confirm findings of the Rodale Research Center (RRC), which indicate that the "maintenance" applications of P and K recommended by many soil test labs are a waste of money. (See "Testing... Testing," *The New Farm*, January, February, and March/April '83.)

"The practice of recommending additional phosphorus and potassium when soil analysis shows high, very high, or excessive levels are already in the soil is wasting farmers' money as well as our limited sources of these fertilizers," says William C. Liebhardt, assistant director of the RRC.

For now, Liebhardt recommends that farmers "Rely on your own judgment and observations. Leave test strips in the fields. Watch for nutrient deficiencies and take tissue samples at the correct times in each field so that you know the nutritional status of the crops.... If your soil test is high and very high, and your tissue tests are adequate, then your plants are telling you they don't need any more nutrients."

Dr. William H. Mitchell

To help make this process easier, RRC staffers are preparing a series of regional handbooks designed to help apply these principles to soil test results.

Mitchell's study includes his 18-year observations on:
- Effects of sub-surface irrigation and fertigation.
- Changes in soil organic matter due to continuous corn cropping and fertilization.
- Micronutrient availability.
- Basing fertilizer and micronutrient decisions on ear leaf analysis.
- Effect of nitrogen fertilizers on soil pH.
- Why application of more than 200 pounds of nitrogen fertilizer per acre fails to increase corn yields.

☐ WHO NEEDS HERBICIDES?/ *It may come as a surprise, but most farmers don't. This special report on the state of the art of mechanical weed control tells why.*

OROSI, Calif.—The farmer on the other end of the phone was desperate. His herbicide had failed, leaving his 200-acre field full of weeds, instead of nice clean rows of sugar beets. His only hope of saving the $125,000 crop was custom cultivator Paul Bezzerides.

"Just how bad is it?" Bezzerides asked.

"You can't really see the rows," the farmer replied.

"That's about as bad as they come," Bezzerides admitted.

Bezzerides had seen worse fields in his nearly half century of building machinery and doing custom cultivating for Western farmers, but not many. "The man's herbicide had failed. He was short of hired labor, and it would have cost him too much—$100 to $200 an acre—to weed the field. He was going to disk it up."

The disk never got out of the implement shed, though. Bezzerides' unique brand of cultivator cleaned up not only the row middles, but weeded the rows, themselves.

Bezzerides believes the application of cold steel is still the best way to control weeds. And more farmers than the chemical companies and the farm press care to admit seem to agree with him.

"Most weed control today is still accomplished non-chemically," according to Dr. David Pimentel of Cornell University. "Mechanical and cultural controls remain the prime means of weed control in this country. Of the 890 million acres of U.S. cropland (including pastures), only 17 percent is treated with herbicides." Take out pastures and the herbicide use figure rises, but only to 34 percent.

$300 Million Market

All that means mechanical weed control is big business. U.S. production of cultivators and weeders in recent years has averaged 60,000 to 70,000 units a year, according to USDA and Commerce Department statistics. In 1981, the most recent year for which figures are available, the value of U.S. cultivator and weeder shipments, including exports, was nearly $300 million.

That's a far cry from the $2.8 billion worth of herbicides sold by U.S. firms in 1981, but the steady demand for mechanical weed control is strong enough to have won a solid commitment from the nation's largest farm machinery manufacturers. It's the lifeblood of literally scores of firms throughout the United States. John Deere & Co., for example, upgrades its line of cultivating and thinning equipment as regularly as it fine-tunes features on new tractor models.

Why? "Because the interest and demand is there," says Randy Zinck, a marketing representative for the Moline, Ill., firm. "We learn what farmers need through market surveys, farm progress shows and individual testing on selected farms across the country."

And what farmers need, especially in these days of soaring production costs and low commodity prices, are safe, efficient and durable products that aren't a constant drain on the pocketbook. Cultivators literally last for decades, while herbicides must be purchased fresh every year. Both systems of weed control have their limitations and drawbacks, but the continuing development of faster and more efficient machinery is making mechanical weed control an integral part of more and more farmers' soil-conserving, less-till systems that are helping regenerate our farmland.

Leading this movement toward more and better mechanical weed control are innovative pioneers like Paul Bezzerides. The 73-year-old Bezzerides never went to college, but his nearly a dozen U.S. patents, and his prosperous custom cultivating, manufacturing and farming businesses are credentials enough.

"Paul's a genius," says Lud Shonnard, supervisor of the 12,329-acre New Columbia Ranch in Firebaugh, Calif. "We use his cultivators exclusively. You can't compare his equipment with something you'd get anywhere else, because it's just not available anywhere else."

Shonnard grows cotton, vegetables, grapes and hay. New Columbia Ranch has bought Bezzerides' cultivators and used his cotton cultivating services since the mid-1970s.

"Most cultivators work up to the row, we work *through* the row," explains Bezzerides. This is accomplished with tools mounted in sequence on the cultivator frame. Running at the front is probably his best-known invention, the spyder.

This offset-spiked wheel runs close to the crop, breaking up and mulching the soil. Spyders may be used to either pull soil away from the row, or turned around to feed loose soil to the plants. Mounted behind the spyders is a set of spring-hoe weeders that cut close to plants. A little soil boils over the narrow hoes, turning soil and killing weeds.

Spinners, spring-loaded rods mounted on a wheel, or thinners, may be mounted to actually run within the row. These reach weeds between the plants, but also thin the crop.

How are all those close-running tools kept on the mark? "The secret is in having a frame that lets you get in close," he says. Under most conditions, Bezzerides runs his frames with self-guidance systems that allow the driver not only

Paul Bezzerides explains how his cultivators work in close to the row.

Bezzerides' spring hoes and spyders (rear view) mounted for normal use.

to run close, but to travel at up to nine miles per hour.

A special, frame-mounted plow cuts a slanted furrow outside the wheel track on each side of the frame. A guide wheel (mounted with a pneumatic tire) rides in this furrow at an opposing angle. This stabilizes and guides the frame during the first cultivation, and serves as a ready-made guide for later cultivations. This system also prevents "crabbing" on slopes.

'Better Ideas' Abound

While Bezzerides has been developing state-of-the-art tools for weed control under local conditions, farmers and small firms in the Midwest and South have been busy building machines for corn, soybeans and other row crops.

Although these innovators have been working independently, their weed control systems have striking similarities. Farmers Francis Boeser, John Davis, and those using Cole Manufacturing Company's mulch/plant system have all achieved outstanding non-chemical weed control. Their programs begin at planting time, *and they all cut a depression, or furrow, just ahead of the planter, rather than planting on the level.*

Florida farmer Davis' furrow is only a few inches deep, while the furrow made by a mulch planter might be half a foot deep. But the concept remains the same. Planting in a furrow helps these farmers *feed soil down to the crop* as it grows, smothering small weeds. This soil takes out the early-sprouting, in-row weeds, which are the toughest to get with a cultivator.

"That's the idea of weed control," says Boeser, who farms 550 acres near Trenton, Ill. "You smother weeds with soil at the right time, rather than trying to cut or pull them out."

Boeser says he rarely has to touch a growing weed with the cultivator. The cultivator is used for *soil control*, rather than ineffective and too-late weed pulling. Boeser gets his corn and beans planted below the level with homemade, adjustable V-plows mounted on his White air planter.

John Deere S-tine cultivators work at speeds up to 7 miles per hour.

Bush Hog has introduced three new cultivator lines to go with its "Vibratine" model (shown here), to allow good mechanical weed control in a variety of soils and crops.

Buffalo slot planter, planting in sod.

Buffalo plateless till planter, working on an old corn row. With the Buffalo system, weed control begins at planting.

The Buffalo all-flex cultivator, moving soil away from the row. The soil will be ridged back to the row later.

These V-plows spread the soil to make a foot-wide furrow. Once the beans are up, but are still very small, he rotary hoes. Since the plants are below level, they are protected from the hoe, which trickles a little soil down to the beans, stopping weeds even at this early stage. When the beans are 4 to 6 inches high, Boeser comes in with the cultivator. He uses a modified, four-row Oliver.

John Davis, who is known for growing certified soybean seed that never tests out at more than .03 percent weed seed, plants in a furrow, too. "Lots of people around here plant beans about on the level, but we plant down in a hole. That way, while the plants are little, you can put all the dirt you want to them when you cultivate."

Davis plants three different soybean varieties after wheat, so the beans don't require cultivation all at one time. His single-ribbed front tires make a clear impression in the soil that is easily followed on later trips across the field.

Mulch & Ridge Tillage

An even more unconventional approach to planting is used with the Cole mulch/plant system. The Cole planter unit deep chisels and parts unprepared soil or stubble with a middlebuster, leaving plenty of soil to feed down to the crop during later cultivations. The cultivator used with this system has a pair of disk hillers that cut in close to the crop to sweep weed seed and grass away from the crop. Single siding sweeps, mounted farther back, undercut weeds and grass at the edge of the furrow and throw clean soil back to the crop.

The Buffalo Till system, by Fleischer Manufacturing Inc., Columbus, Neb., also features a planter that works in stubble or unprepared soil. Proper weed control with this system involves: planting into a ridge made the previous year; using disk hillers to push soil away from the crop on the first cultivation; and reversing disk hillers to push soil back to the crop, forming a new ridge during later cultivations.

"When the corn is 12 to 18 inches high or so, I come through with a cultivator that takes the loose soil and residue between the rows and pushes it up into the corn to form a ridge. If there are any small weeds in the row, they get covered up in the process," says Iowa farmer Ernest E. Behn, author of "More Profit With Less Tillage", a book about the Buffalo system.

And with both the Cole and Buffalo methods, crop residues are kept near the soil surface where they protect soil from erosion and are turned into humus by bacterial action.

Herbicides, on the other hand, can harm the soil. "The potentially disruptive effects of herbicides in the agro-ecosystem include: increased incidence of disease and insect pest damage of crop plants, population disturbance of soil microbial flora and fauna, and disruption of natural enemy populations," according to a report by Sean L. Swezey, of the University of California, and Edwin J. McLeod, of the Organic Agriculture Research Institute, Graton, Calif. "In addition, the selective pressure of herbicides has caused some weed species to become resistant to herbicides, thus warranting other means of control."

Cultivation—the primary means of weed control in this country.

Self-Guiding Machines

While most agricultural equipment companies have not developed the types of weed control systems being explored by firms like Cole and Buffalo, they have continued to improve the basic cultivating equipment for improved reliability, strength, and ease of use.

For example, J&J Guide Systems Inc., Sanborn, Minn., offers a complete guide system for precision cultivating. Rather than having the cultivator frame alone fitted with a guide system, as with Bezzerides' equipment, J&J has both the tractor's front wheels and the cultivator frame following a groove. This groove is made in the row middles during planting.

"You first mount cultiguides (soil-cutting shoes) on the planter. They're positioned on the planter exactly where the front tires of your tractor will be riding," says Greg Gilb, of J&J Guide. "When you're ready to cultivate, you mount single-ribbed tires, which follow this groove. You'll literally be able to let go of the steering wheel and still have the tractor hold steady. You'll be doing a precision cultivating job because the cultivator is right there.

Bourquin weed puller plucks out weeds that would normally require hand rogueing or chemical applications.

"By having the cultiguides mounted on the cultivator, they are going to set down in their original tracks that were made at planting time. When the cultiguides are in the tracks, they don't allow the cultivator to slide into the rows. You can set your shovels much closer to the row than you normally can. This is especially true in narrow-row soybeans."

A Weed-Wringer

Sometimes a few large weeds still tower above crops, despite earlier cultivations. Hand rogueing or selective chemical application used to be the only choices. But now, Dan Bourquin, of Bourquin Design and Manufacturing in Colby, Kan., has developed a machine that brushes along the top of a crop and actually pulls weeds in the row. "A series of wheels, which are driven hydraulically, rotate like a wringer on an old wringer washer, grasping weeds which extend above the crop row, pulling them upward and out," he says.

The weed puller may be front- or rear-mounted. It has been used for removing shattercane from soybeans and grain sorghum; careless weed from cotton; volunteer corn; coffeeweed, beggarweed and pigweed from soybeans; and various other weeds from sugar beets and vegetables.

H.L. Walser, a cotton farmer in Seagraves, Texas, says the weed puller can be used any time weeds grow 8 to 12 inches above the crop. "It just hauls them up out of the ground. You see the results immediately, behind the machine.

"Once you get in the field, you can just keep going. You don't have to stop every so often to fill up on $200 to $300 worth of chemicals," says Walser. Is the weed puller more economical than the herbicides Walser was using? "You figure it out," he says. "The machine's been used on 2,500 acres and it looks like new."

John Deere & Co. has introduced several modifications to its row crop cultivating and thinning equipment lineup. "The major change we've made is that we are now using an 'on the square' rig design that features square rig beams and cross-arms that lock the shanks in line for accurate cultivation," says Deere's Randy Zinck. The idea of this construction is to prevent side sway.

Another Deere addition is a precision rig for vegetable crops and sugar beets. "This rig has tapered roller bearings where the lower rig links connect to the coupler plates and main-frame," Zinck says.

What may turn out to be Deere's most popular innovation is the new handcrank-adjustable gauge wheel. This crank replaces the time-consuming, knuckle-bearing bolt type adjuster. Deere also offers two- and four-row cultivators for small fields and tractors.

With the millions of dollars spent each year on herbicide advertising—Lasso ads on television, and seven-page color spreads for Dual in most farm magazines—it's often easy to overlook mechanical weed control. But, as Cornell's Pimentel points out, "Even when growers use herbicides, they frequently also employ mechanical and cultural controls."

There's one, simple reason for that. Just ask any farmer like Frances Boeser, who uses no herbicides on his own farm, but treats rented acreage with chemicals his landlord insists on.

"It's just more profitable without the chemicals," he says.

☐ GROW LESS AND MAKE MORE

Suppose every farmer in the United States suddenly quit using synthetic fertilizers and chemical pesticides.

Popular belief has it that they'd go broke overnight; that Americans and countless people throughout the world would soon starve.

But that's not the way two Iowa State University researchers see it. The bottom line, according to a 1979 computer projection by Drs. Kent D. Olson and Earl O. Heady, is that farmers would make more money farming organically. A lot more money.

Instead of the $5.4 billion in farm income Olson and Heady projected for seven field crops and seven livestock enterprises for 1980 under a continuation of existing farming and export practices, the income from nationwide organic farming would have been $21.6 billion—a nearly fourfold increase. Expanding exports under conventional methods to use almost all available U.S. cropland, the second of three possibilities in their computer model, would have earned farmers only $6.23 billion.

"The nation could readily meet domestic demands under the organic farming alternative, but the export potential is decreased considerably," Olson and Heady report. They estimate conventional farming would have produced 2.3 billion bushels each of wheat and corn for export in 1980, compared with an export potential of 400 million to 500 million bushels of each grain with organic methods.

Considering the grim economic picture facing farmers, lower yields and fewer exports might be just what we need. Farmers now are growing more than ever. But, with high interest rates, record land prices, some of the worst soil erosion in history, and steady inflation for machinery, fuel, fertilizer and chemicals, it's also costing them more than ever. They're growing more and getting less for it. Corrected for inflation, net farm income is about the lowest it's been since the Great Depression.

"Production costs in just one year have increased 20 percent for corn and 16 percent for wheat, and prices are down at least 10 percent," says George Casler, an agricultural economist at Cornell University. "All in all, the record (1981) crops should benefit almost everyone somewhat, except those who toiled to produce them—the farmers."

Some would argue that our $43.8 billion worth of farm exports in 1981 is the only thing that kept that year's roughly $30 billion foreign trade deficit from being more than twice as bad. But if we didn't buy so much foreign oil for gasoline and diesel fuel, plastics and pesticides, and increasing amounts of foreign natural gas for nitrogen fertilizer, then we wouldn't have such a big trade deficit in the first place. We wouldn't have to cut down shelter belts, plow up terraces and pastures, and trade soil for oil.

Reducing U.S. farm exports also would force other countries, especially those in the Third World, to become more self-sufficient in producing the basic foods they need to feed burgeoning populations, instead of depending on imported foods and expensive, non-renewable technologies.

But the export picture may not be as bleak with organic farming as Olson and Heady predict. Their study projects organic yields at less than half those under conventional methods. For example, they see organic yields of 20 bushels of wheat per acre; 49-bushel corn; 20-bushel soybeans; and 17-bushel yields on other feed grains.

If Olson and Heady had traveled a few miles west of Ames, where Iowa State is located, they would have found Richard Thompson who consistently produces 125-bushel corn organically. Then there's Del Akerlund, an organic farmer in Valley, Neb., who grows 150-bushel corn, 55-bushel soybeans and 100-bushel oats. Or Steven Garnett with his 140-bushel organic corn in Remington, Va. The list goes on and on.

"The study does not attempt to select 'a best' farming alternative," Olson and Heady say in their report. "Rather, it estimates some expected effects on land use, supply prices...

"We have not attempted to answer questions concerning the value of environmental impacts associated with a change in farm technology. These questions also need to be answered before the U.S. public can make reasonable choices for the future.

"The analysis indicated that the organic farming alternative... could meet domestic demand and low levels of exports. This would be accomplished with considerably higher food prices... Total farm income also would increase in all regions. The United States would not have a large reserve of land ready to be tapped to 'feed the world' under recurrence of a crop shortfall elsewhere. These are some of the trade-offs involved, but only the people of the United States and the world can appropriately weigh these trade-offs."

For U.S. farmers caught in a suffocating cost-price squeeze, the choice should be clear.

Appendix

REGENERATIVE AGRICULTURE RESOURCE LIST

Most are producer groups, but some are purely academic, research-oriented, or both. All should be good sources of information, however.

U.S. NATIONAL

Institute for Alternative Agriculture (1982)
9200 Edmonston Rd., Suite 117
Greenbelt, Md. 20770
(301) 441-8777, 8 a.m. to 5 p.m.

Members Produce: A variety of crops, livestock.

Washington-based clearinghouse on alternative agriculture, a "voice" to support research and education. Publishes monthly newsletter, organizes annual policy symposium, plans to publish refereed scientific journal on alternative agriculture by 1986.

For membership contact: office at above address.

Membership fee: $15/year.

Regenerative Agriculture Association (1982)
222 Main St.
Emmaus, Pa. 18049
(215) 967-5171

Members Produce: Wide range of crops, livestock.

Publishes *The New Farm* magazine, books and other special publications, sponsors field days; operates readers' service that finds answers to farmers' specific questions or makes referrals to ag researchers or other farmers via *Farmers Own Network for Extension;* dedicated to putting people, profit and biological permanence back into agriculture. Affiliated with the Rodale Research Center, R.D. 1, Box 323, Kutztown, Pa. 19530.

For membership contact: James Morgan, executive director, at above address.

Membership Fee: $15/year.

CALIFORNIA

California Agrarian Action Project (1977)
433 Russell Blvd.
Davis, Calif. 95616
(916) 756-8518

Members Produce: Wide variety of crops.

Publishes newsletter, provides pesticide safety training for farmworkers, sponsors seminars (IPM, economics, etc.), produces annual market directories for the organic food industry. Local chapters in Yolo, Fresno, Monterey counties.

For membership contact: office manager at above address.

Membership fee: $15/year.

California Certified Organic Growers (1972)
1920 Maciel Ave.
Santa Cruz, Calif. 95062
(408) 476-0504

Members Produce: Field crops, vegetables, fruits.

Operates organic certification program, coordinates marketing produce. Local chapters include:

Fresno-Tulare Chapter
P. Freestone, secretary
Freestone Farms
42126 Road 168
Orosi, Calif. 93647
(209) 528-3816

South Coast Chapter
Wendy Krupnick, secretary
1383 Sycamore Canyon Rd.
Santa Barbara, Calif. 93108
(805) 966-1012

Butte-Glenn Counties Chapter
W. Schwartz, secretary
Rt. 2, Box 2450
Oreland, Calif. 95963
(916) 865-2015

Yolo Chapter
George Stevens, secretary
P.O. Box 107
Capay, Calif. 95607

Central Coast Chapter
P.O. Box 1143
Freedom, Calif. 95019
(408) 726-3100

Big Valley Chapter
Malia Sheldon, secretary
1007 College Ave.
Modesto, Calif. 95350
(209) 521-2816

Mendocino Chapter
P.O. Box 246
Talmage, Calif. 95481

North Coast Chapter
Sy Weisman, secretary
407 Furlong Rd.
Sebastopol, Calif. 95472
(707) 823-0650

For membership contact: main address above.

Membership fee: $25/year, growers; $10/year, businesses; $10/year, supporting members.

INDIANA

Indiana Organic Growers Association (1980)
Paul Schellenberger
Rt. 4, Box 278
Floyd Knobs, Ind. 47119
(812) 923-3689

Members Produce: Vegetables and herbs.

Holds quarterly meetings, conducts farm tours, promotes organic farming, publishes quarterly newsletter, *Hoosier Organic Farmer.*

For membership contact: Paul Schellenberger at above address.

Membership fee: $6/year.

KANSAS

Kansas Organic Producers (1975)
c/o Raymond Meyers
Rt. 2, Box 45
Lincoln, Kan. 67455
(913) 524-4578

Members Produce: Variety of crops.

Publishes newsletter, conducts farm tours, holds annual conference.

For membership contact: Ray Meyers at above address.

Membership fee: $25 for organic growers; $15 supporting.

KENTUCKY

Kentucky New Farm Coalition (1978)
James D. Dixon, president
Rt. 2, Box 66
Hustonville, Ky. 40437
(606) 787-9869

Members Produce: Wide variety of crops.

Publishes quarterly newsletter, *The Kentucky New Farm Gazette;* holds annual winter meeting; supports development of a regionalized, ecological agriculture and preservation of the small family farm lifestyle.

For membership contact: Bud Blackwell, 2013 Edgeland Ave., Apt. 1, Louisville, Ky. 40204.

Membership fee: $12/year.

MICHIGAN

Organic Growers of Michigan (1972)
Rick Pahl, president
Rt. 2, Box 102
Lawrence, Mich. 49064
(616) 674-3474

Members Produce: Grains, dairy, fruit, vegetables.

Active in group marketing and purchasing, promotion of organic produce. Most chapters hold monthly meetings. Active chapters include:

Thumb Chapter
Lewis King
3031 White Creek Rd.
Kingston, Mich. 45741
(517) 683-2541

Southwest Chapter
Rick Pahl
Rt. 2, Box 102
Lawrence, Mich. 49064
(616) 674-3474

Raisin Valley Chapter
Larry Boyd
12345 Lab Rd.
Brooklyn, Mich. 49230
(517) 592-8552

Thornapple Valley Chapter
Randy Bond
11840 Potters Rd.
Lowell, Mich. 49331
(616) 897-5762

Membership fee: $10/year.

MINNESOTA

Minnesota Organic Growers and Buyers Association (1972)
Box 9747
Minneapolis, Minn. 55440
(612) 674-8527

Members Produce: Field crops, small grains, corn, dairy, poultry, beef, vegetables.

Supports certification standards program; holds educational conferences; conducts farm tours; supports marketing; provides limited technical information for growers in Minnesota, western Wisconsin, eastern North and South Dakota.

For membership contact: Yvonne Buckley at above address.
Membership fee: $20/year.

MISSISSIPPI

Mississippi Organic Growers Association (1980)
800 Colonial Circle
Jackson, Miss. 39211

Members Produce: Vegetables and fruits.

Publishes newsletter, holds annual seminar. Southeast Chapter holds monthly meetings, arranges bulk purchase of natural minerals, supports certification. Also has Metro Jackson Chapter.

For membership contact: Don West, treasurer, at above address.

Membership fee: $5/year.

MISSOURI

Ozark Permaculture Society (1982)
722 Cliff St.
Jefferson City, Mo. 65101
(314) 634-5557

Members Produce: Extremely varied; interested in plants for good land use and soil fertility.

Organizes occasional conferences, publishes *Ladybug* newsletter, arranges farm apprenticeships, operates lending library, supportive of marketing produce.

For membership contact: Mary Lehmann at above address.

Membership fee: None.

MONTANA

Alternative Energy Resources Organization (1974)
324 Fuller, C-4
Helena, Mont. 59601
(406) 443-7272

Members Produce: Small grains, hay.
Publishes *AERO Sun-Times* bi-monthly, sponsors conferences, promotes sustainable farming methods and systems of agricultural and economic development.

For membership contact: Beck Newell, office manager, at above address.

NEBRASKA

Nebraska Organic Agriculture Association (1977)
Rt. 1, Box 163
Marquette, Neb. 68854
(402) 854-3165

Members Produce: Field crops and livestock.
Organizes annual meeting/workshop, publishes newsletter. Most active in eastern Nebraska and nearby counties of surrounding states.

For membership contact: Robert Staffen, Rt. 1, Box 110, Bennington, Neb. 68007.

Membership fee: $25/year, producer; $10/year, consumer; $4/year, newsletter; $100/year, supporting.

Small Farm Resources Project (1976)
P.O. Box 736
Hartington, Neb. 68739
(402) 254-6893, weekdays 8 a.m. to 5 p.m.

Not a membership organization but issues newsletters and various literature through subscriptions. Supports political action; operates lending library; sponsors internships, on-farm research for dryland agriculture.

Affiliations:
Center for Rural Affairs
P.O. Box 405
Walthill, Neb. 68067

NEW ENGLAND

Maine Organic Farmers and Gardeners Association (1974)
283 Water St.
Augusta, Maine 04330
(207) 622-3118

Members Produce: Field crops, vegetables, fruits.

Organizes conferences/workshops, publishes bi-monthly newsletter, arranges farm apprenticeships, operates lending library, supports marketing, runs the annual Common Ground Country Fair (40,000 attendance). Chapters function in the Bradford area: Jeff and Susan Parker, (207) 327-1246; Mid-York County area: (207) 676-2209; Somerset County area: Bill Rowe, RFD 1, Box 664, Madison, Maine 04950.

For membership contact: main office listed above.

Membership fee: $15/year.

Natural Organic Farmers' Association
Members produce wide variety of crops. NOFA sponsors periodic regional and state meetings, conferences, farm tours; members receive the quarterly NOFA *The Natural Farmer* as well as state newsletters. Some chapters form bulk-buying groups, coordinate apprenticeship programs, support organic certification. Membership fee in all states is $15/year.

NOFA/Connecticut (1982)
670 Wintergreen Ave.
Hamden, Conn. 06415
(603) 789-7865

For membership contact: Kathleen Mulligan, 100 Rose Hill Rd., Branford, Conn. 06405
(603) 681-0639

NOFA/New Hampshire (1970)
P.O. Box 335
Antrium, N.H. 03440

For membership contact: June Francis at above address.

NOFA/ New York (1981)
P.O. Box 454
Ithaca, N.Y. 14851
(315) 475-7230

For membership contact: address above.

NOFA/Massachusetts (1981)
21 Great Plain Ave.
Wellesley, Mass. 02181
(617) 235-1447

For membership contact: Stacy Miller at above address.

NOFA/Vermont (1971)
43 State St.
Montpelier, Vt. 05602

For membership contact: Wendy Cole, Quail John Rd., E. Thetford, Vt. 05043.

The New Alchemy Institute (1969)
237 Hatchville Rd.
E. Falmouth, Mass. 02536
(617) 563-2655

Members Produce: Mostly vegetables

Publishes newsletter; arranges farm apprenticeships; supports marketing; sponsors courses on solar greenhouse management, permaculture, tree crops.

For membership contact: Kim Allsup at above address.

Membership fee: $35/year.

The New England Small Farm Institute (1978)
Jepson House, Jackson St.
Box 937
Belchertown, Mass. 01007
(413) 323-4531, weekdays 9 a.m. to 5 p.m.

Teaches students to operate a successful, diversified organic farm. Maintains a library for visitors (does not lend out materials), a two-acre demonstration site; leads tours and arranges informational meetings. Program developed through state food and agriculture department.

Woods End Agricultural Institute
Orchardhill Rd.
RR Box 128
Temple, Maine 04984
(207) 778-9241

Not a membership organization. Conducts agricultural research stemming from work of the Woods End (soil testing) Laboratory, RFD Box 65, Temple, Maine 04984.

NORTH DAKOTA

North Dakota Natural Farmers Association
C/O Kirschenman Family Farms
Windsor, N.D. 58493
(701) 763-6287

Members Produce: Small grains, row crops. Publishes periodic newsletter, sponsors annual convention, helps market produce.

For membership contact: Fred Kirschenman at above address.

Membership fee: $25/year.

OHIO

Firelands Organic Producers Association (1981)
RR 2, 1716 Remelle Rd.
Monroeville, Ohio 44847

Members Produce: Field crops, fruit, vegetables.

Holds local meetings for discussions, lectures, films. Active primarily in Huron and Erie counties in north-central Ohio.

For membership contact: Jim Hemminger at above address.
Membership fee: None.

Ohio Ecological Food and Farm Association (1979)
7300 Bagley Rd.
Mt. Perry, Ohio 43760
(614) 448-6545

Members Produce: Wide range of crops.

Publishes newsletter and directory, conducts farm tours, supports certification program, holds annual all-day workshop conference.

For membership contact: Ed Kruse, Rt. 3, Box 466, Glouster, Ohio 45732.

Membership fee: $10/year, individual; $24/year, family; $50/year, organization.

PACIFIC NORTHWEST

Tilth (1974)
4649 Sunnyside North
Seattle, Wash. 98103
(206) 633-0451, Mon. through Thurs. 9 a.m. to 2 p.m.

Members Produce: Fruits, berries, vegetables.

Publishes *Tilth* quarterly, arranges local farmers' markets, holds periodic meetings, sponsors demonstrations, studies farm policy, operates lending library. Covers Washington, Oregon, Idaho, western Montana, northern California and British Columbia. Local chapters in all but Montana and British Columbia.

For membership contact: main office at above address.
Membership fee: $12/year.

PENNSYLVANIA

Bio-Dynamic Association (Southeast Pennsylvania Branch) (1978)
RD 3, Box 230
Phoenixville, Pa. 19460

Members Produce: Dairy, bread products.

Organizes conferences/seminars, publishes newsletter, arranges farm apprenticeships for the Southeast region of the state. Affiliated with the Bio-Dynamic Farming and Gardening Association, Richmond Townhouse Rd., Wyoming, R.I. 02898.

For membership contact: Roderick Shouldice at above address.
Membership fee: $15/year.

Western Pennsylvania Organic Growers (1982)
130 Critchlow Rd.
Renfrew, Pa. 16053
(412) 586-7683

Members Produce: A variety of crops.
For membership contact: Pat McKinney at above address.
Membership fee: $10/year.

SOUTH

Carolina Farm Stewardship Association (1980)
P.O. Box 205
Bynum, N.C. 27228
(919) 542-5029 (evenings)

Members Produce: Mostly fruits and vegetables, some livestock. More than half are gardeners.
Organizes conferences, publishes newsletter, operates lending library, issues marketing directory, developing a certification program.

For membership contact: Mandy Kinney, Rt. 1, Box 132, Danbury, N.C. 27016.
Membership fee: $12/year.

Tennessee Alternative Growers Association (1980)
Rt. 6, Box 526
Crossville, Tenn. 38555

Members Produce: Vegetables, fruits, field crops, specialty crops.

Organizes conferences, publishes newsletter, supports political action, markets produce, conducts some research. Several chapters meet periodically.

For membership contact: office at above address.

Membership Fee: $10/year.

Virginia Association of Biological Farmers (1974)
Box 252
Flint Hill, Va. 22627
(703) 675-3263

Members Produce: Small livestock, small grains, vegetables.

Conducts several meetings a year, including a day-long program focusing on various concerns (such as IPM, marketing, use of certain products, soil building). Publishes quarterly newsletter.

For membership contact: membership chairman at above address.

Membership fee: $7.50/year.

SOUTH DAKOTA

Soil Association of South Dakota (1976)
Rt. 1, Box 132
Tripp, S.D. 57376
(605) 935-6044

Members Produce: Field crops.

Publishes newsletter, acts as support group, educates members and public.

For membership contact: Linda R.W. Schnabel at above address.

Membership fee: $10/year.

CANADA NATIONAL

Canadian Organic Producers Marketing Cooperative Ltd. (1983)
Alfred Moore, president
Box 118
Dinsmore, Saskatchewan S0L 0T0
(306) 846-4611

Members Produce: Primarily grains, livestock, vegetables. Active in marketing, following growers' own organic standards; promotes non-chemical government farm policy by keeping agencies aware of advances in organic technology. (A project of the Saskatchewan-based Back-to-the-Farm Foundation, a non-profit provincial group formed in 1974 to find the best ways to produce pure food and get more people onto the farm. Foundation headed by Elmer Laird, Box 69, Davidson, Saskatchewan S0G 1A0. Membership fee is $10/year.)

For co-op membership contact: Alfred Moore, president, at above address.

Membership fee: $50 for active member (selling produce); $25, associate member.

Canadian Organic Growers (1978)
33 Karnwood Drive
Scarborough, Ontario M1L 2Z4

Members Produce: Field crops, vegetables, fruits, meats, eggs, herbs, honey, etc. Highly diversified, but most members are gardeners.

Organizes conferences; publishes quarterly newsletter; operates a mail-order lending library; supports political action, produce marketing, organic certification.

For membership contact: Lida McMartin, 146 Elvaston Drive, Toronto, Ontario M4A 1N6.

Membership fee: $5/year.

MARITIME CANADA

Maritime Sustainable Agriculture Network (1983)
c/o Jane Kehoe
RR 1
Wolfville, Nova Scotia B0P 1X0
(902) 542-2857

Members Produce: Wide variety of crops.

An umbrella group supporting the small Atlantic province farmer interested in sustainable methods of maintaining soils, combating insects and viruses, growing crops and marketing goods. Sponsors conferences and networks for students and organized groups, building a reference/resource library, distributes organic farm supplies and *Between the Issues* tabloid.

For membership contact: Jane Kehoe at above address.

Membership fee: $10/year.

ONTARIO

Ecological Farmers of Ontario (1970)
RR 1
Tiverton, Ontario N0G 2T0
(519) 368-7417

Members Produce: Wide range of crops.

Holds annual spring and fall conferences for members across Ontario. Also holds spring introductory seminar, conducts summer farm tours, publishes quarterly newsletter. (Formerly the Natural Farmers Association of Ontario.)

For membership contact: Lawrence Andres, chairman, at above address.

Membership fee: $15/year.

PRINCE EDWARD ISLAND

Prince Edward Island Organic Growers Association (1984)
RR 3
Hunter River
Prince Edward Island C0A 1N0
(902) 964-3041

Members Produce: dairy, small grains.

Supports certification, marketing efforts; involved in promoting organic produce.

For membership contact: Alan Hearn at above address.

Membership fee: $7.50/year.

FOR FURTHER READING

The books, periodicals and magazine articles on this list should be sitting on the shelves of most public or college libraries. Once you locate the materials, you're likely to find related titles nearby. The basic principles of soil fertility, crop rotations and manure application apply equally well to all types of agriculture. Doing a little homework will help answer many basic questions, and such reading will suggest a slew of new questions, answers and ideas that will help make your farm more profitable now, and for many years to come.

COMPOSTING

"Composted Manure Cutting Fertilizer Costs" by Roy Alleman. *The New Farm,* July/Aug. 1982, p. 12.
Describes several Midwest feedlot operations and farmers who are turning their large manure wastes into compost.

"Waste Not, Want Not" by Dr. Leon Chesnin. *The New Farm,* March/April 1983, p. 12.
Describes benefits of large-scale composting operations in eight Nebraska cities and on a growing number of the state's farms.

"The Rodale Guide To Composting" by Jerry Minnich and Marjorie Hunt, 1979.
Explains how a compost pile works; shows how to build several types, and maintain a farm-scale pile. Available for $14.95 from Rodale Press, Emmaus, Pa. 18049.

"Composting Of Farm Manure" Small Farm Energy Project report no. 8, Jan. 1980.
Explains the composting process and shows the needs of the working pile. Also details home-built and commercial turners, includes simplified diagrams. Available from Small Farm Energy Project, P.O. Box 736, Hartington, Neb. 68739.

COVER CROPS/GREEN MANURES

"Winter Annual Cover Crops For Non-Tillage Corn Production" by W.H. Mitchell and M.R. Teel. *Agronomy Journal,* Vol. 69, 1977, p. 569-572.

"Leguminous Crops For Green Manuring" by C.V. Piper. How to select and use green manures, including Canadian peas, cowpeas, soybeans, velvet beans, clovers, vetches. USDA Farmer's Bulletin No. 278, 1907.

"Green Manures—A Mini-Manual" published Jan. 1983 by the Research Department of Johnny's Selected Seeds.
Provides general background on the uses and advantages of green manuring, including how to plant, care for, and till both legumes and non-legumes. Available for $1 from Johnny's Selected Seeds, Box 2580, Albion, Maine 04910.

Cover Crop Series by William K. Kruesi.
Focuses on the value and use of winter rye, buckwheat and red clover as cover crops. Available for 25 cents each (postpaid) from University of Vermont Extension, 31 The Green, Woodstock, Vt. 05091.

"Feed The Soil" by Edwin McLeod. Published in 1982 by Organic Agriculture Research Institute.
Contains a useful 75-page section detailing growth, use, ranges and seeding instructions for virtually every green manure crop used in America. Also has good planning charts, soil climate key. E.B. White would like this book. Talking rabbits introduce basic natural farming principles. Available from the Organic Agriculture Research Institute, P.O. Box 475, Graton, Calif. 95444.

"Follow-Up On Fava Beans" by Corwin Rife and John Yaeger. *The New Farm,* March/April 1981, p. 28.
Helps determine whether this cold-hardy legume cover crop can fit into your rotation.

"Science Is Rediscovering Green Manures" by Jack Sperbeck. *The New Farm,* July/Aug. 1981, p. 50.
Describes research on use of legumes in crop rotations to help control pests and produce nitrogen.

"Green Manuring Principles And Practice" William Brinton's translation of Otto Schnid's and Reudi Klay's 40-page booklet.
Summarizes recent research on green manures in Switzerland and includes tables comparing characteristics of selected green manures, their nutrient values. Also compares suitability of various machinery for tilling green manures. Available for $5 from the Woods End Agricultural Institute, RR Box 128, Temple, Maine 04984.

"Overseeding Research Results: 1982-1984" by Robert Hofstetter.
Rodale Research Center agronomist studies promising legume cover crops seeded in corn and soybeans to determine effects of planting time and seeding rate, winter survival, and nutrient inputs. Report No. RRC/AG-84/29, available for $2 from the Rodale Research Center, RD 1, Box 323, Kutztown, Pa. 19530.

CROP ROTATIONS

"Fitting The Right Legume To Your Crop Rotation System" by Gene Logsdon. *The New Farm,* Nov./Dec. 1980, p. 44.
Details many legume varieties, along with their best uses, characteristics, strengths and weaknesses in various regions.

"The Rotation Effect—What Causes It To Boost Yields?" by R. Kent Crookston. *Crops and Soils,* March 1984, p. 12.
University of Minnesota agronomist argues that farmers can increase net returns by exploiting the still unexplained higher yields potential of crop rotations.

"**A Cash And Cover Crop**" by Mike Brusko. *The New Farm,* Sept./ Oct. 1984, p. 21.
Lana woolypod vetch eliminates the need for purchased N in no-till rice on 900-acre California rice farm.

"**He Cut His Chemical Bill In Half**" by Mike Brusko. *The New Farm,* Sept./Oct. 1984, p. 26.
Florida seed grower uses a chisel plow and strict crop rotation to keep his soil in top shape and reduce his need for fertilizers and herbicides.

"**Planting The Seeds Of Profit**" by Mike Brusko. *The New Farm,* Nov./Dec. 1984, p. 30.
Illinois corn and soybeans farmer discovers the nutrient and yield advantages of legumes in crop rotations that save soil and cut fertilizer and pesticide costs.

ECONOMICS

"**The Kutztown Farm Report**" by Martin N. Culik, et al., 1983.
Rodale Research Center agronomists report on five-year study of a low-input crop/livestock farm in southeastern Pennsylvania that produces good yields at 10 percent less cost than comparable farms. Available for $5.95 from the Regenerative Agriculture Association, 222 Main St., Emmaus, Pa. 18049.

"**Economics Of Alternative Crop Rotation Systems For East-Central Nebraska—A Preliminary Analysis**" by Glenn A. Helmers, et al.
The production costs and returns of crop rotations prove competitive with conventional continuous cropping, but organic rotations' net returns peak at $125/acre ($150/acre with straw sales), compared with $94/acre in continuous corn. Available free from the Department of Agricultural Economics, University of Nebraska, Lincoln, Neb. 68508.

"**The Macro Implications Of A Complete Transformation Of U.S. Agricultural Production To Organic Farming Practices**" by James A. Langley, et al., 1982.
Ag economists project farm prices if all U.S. farmers suddenly quit using synthetic fertilizers and pesticides: Farmers would produce less but would make a lot more money. Available free from the Center for Agricultural and Rural Development, 578 East Hall, Iowa State University, Ames, Iowa 50011.

INSECT CONTROL

"**The CRC Handbook Of Pest Management In Agriculture**" edited by David Pimentel, published 1981 by CRC Press, Boca Raton, Fla.
Hefty, three volume-compilation of papers on biological pest management.

"**Outsmarting The Onion Maggot**" by Mike Brusko. *The New Farm,* Sept./Oct. 1984, p. 30.
Highlights a study by two Michigan State University entomologists showing that foliar sprays for onion maggot control are a waste of money. Offers non-chemical alternatives.

The IPM Practitioner published 11 times a year by the non-profit Bio-Integral Resource Center, $25/year. Available from the BIRC, 1307 Acton St., Berkeley, Calif. 94706.

"**Plants Pests Don't Like**" by Mike Brusko. *The New Farm,* May/June 1984, p. 22.
Insect-resistant plants are starting to offer farmers a solid substitute for pesticides.

"**Nature Bats Last**" by Mike Brusko. *The New Farm,* May/June 1984, p. 16.
Outlines how the overuse of pesticides ultimately leads to resistant strains of insects that will require different, and perhaps more toxic, chemicals to kill.

"**These Farmers Grow Their Own Insecticides**" by Joel Grossman. *The New Farm,* May/June 1982, p. 28.
California farmers are curbing rising insecticide costs with use of homegrown beneficial insects.

"**A Menu For Murder**" by Gene Logsdon. *The New Farm,* Sept./Oct. 1982, p. 40.
The planting of snap and lima beans provides a good trap crop for the Mexican Bean Beetle, a pest of the Midwest's soybean crop. After the trap crop has done its job, it can be easily disked and/or sprayed. The beetles can then be subjected to their natural enemies without widespread spraying of the main soybean crop.

"**Rodale's Color Handbook Of Garden Insects**" by Anna Carr.
Garden-oriented but good background. Includes 300 full-color photographs of pests common to field and garden crops. Available for $10.90 from Rodale Press, Emmaus, Pa. 18049.

MANURE

"**How Manure Affects Corn Yields**" by Stuart D. Klausner and D. Wilson. *The New Farm,* January 1981, p. 61.
Outlines New York research on the potential of dairy manure as a nitrogen source for corn production.

"**Just How Good Is Your Manure?**" by Vaughn Holyoke. *The New Farm,* Nov./Dec. 1982, p. 14.
Vermont Extension agent offers advice on how a farmer can evaluate and apply manure, as well as its effect on soil tilth.

"**Using Manure Resources Wisely**" by editors of *The New Farm*, 1981.
Compilation of several Extension reports on manure handling and application. Includes nutrient values, diagrams, cost comparisons of different management systems, application rates and incorporation. Available for $2 from the Regenerative Agriculture Association, 222 Main St., Emmaus, Pa. 18049.

"**Managing Animal Manure As A Resource, Basic Principles, Parts I and II**" by Stuart Klausner and David Bouldin, 1984.
Cornell University agronomists discuss manure nutrient values, nitrogen forms and their effects on crop yields, and manure use in a soil fertility program. Part II focuses on

land application, complete with tables and worksheets for calculation. Geared for New York's growing conditions, but contains good background for all farmers with manure. Available for 35 and 40 cents, respectively, from Media Services Distribution Center, 7 Research Park, Cornell University, Ithaca, N.Y. 14850.

"**Unlocking Manure's Magic**" by Mike Brusko. *The New Farm,* January 1985, p. 26.
Summarizes manure-handling tips by Klausner and Bouldin (see above reference). Provides updated versions of tables and worksheets.

"**Solid Manure Handling For Livestock Housing, Feeding, And Yard Facilities In Wisconsin**" by E.G. Burns and J.W. Crowley.
Outlines how to choose the best handling system; describes run-off control structures, and how to plan a solid manure system. Gives examples for a 60-, 75- and 125-head dairy farm, and a poultry farm with 10,000 caged birds; covers construction, loading equipment, fly control, soil fertility values. Extensive appendix and drawings for different plans for storage, stackers, housing layouts, plus an extensive publications list on other specifics. Now out of print but possibly available in local ag-oriented library. Published by University of Wisconsin, 1972.

"**Managing Animal Wastes—Guidelines For Decision-making**" by USDA's Economic Research Service, 1981.
Discusses economic considerations of farms evaluating changes in waste management systems. Examines six different farms (dairy, hog, feeder beef) and estimates their costs and management options, returns, labor requirements. Report No. ERS-671, available for $10 from U.S. Government Printing Office, Superintendent of Documents, Washington, D.C. 20402.

"**Improved Manure Handling Put This Dairy Back On Track**" by Earl F. Spencer. *The New Farm,* Sept./Oct. 1980, p. 35.
New York dairyman tells how he uses a 40-foot stacker and concrete pit to change cow manure from a problem to a resource.

"**Manures And Compost**" by A.J. MacLean and F.R. Hare. Publication No. 868 from Agriculture Canada, 1979.
Describes the benefits of various animal manures, handling methods, green manures and composting. Available from Agriculture Canada Information Services, Ottawa, Canada K1A 0C7.

"**The Manure Primer**" by Winston Way, May 1983.
Vermont Extension agronomist discusses manure's composition, management problems, application, benefits. Also includes a management checklist. Available from Publications Office, Extension Service, Morrill Hall, University of Vermont, Burlington, Vt. 05405.

The New England Farmer publishes its annual "**Manure Special**" every October.
Covers the basics and specifics about the uses and handling of manure. Yearly subscription is $10, available from *NEF,* Box 391, 50 Bay St., St. Johnsbury, Vt. 05819.

ON-FARM RESEARCH

"**Design Your Own Test Plot**" by Victor Wegrzyn and Charles Kauffman. *The New Farm,* February 1981, p. 22.
Tells how a farmer can test new ideas and products by first trying them on a small scale. Shows how to choose a good test plot and carefully harvest for accurate yield figures before drawing a conclusion.

"**Improving Your Gardening With Backyard Research**" by Lois Levitan, 1980.
Cornell University-trained agronomist describes how to design and carry out useful agricultural experiments. Although developed for the home gardener, the book draws on agricultural research and includes farm-scale examples. Now out of print. Published by Rodale Press Inc., Emmaus, Pa. 18049.

"**Big-Time Farmer Research**" by Rex Gogerty. *The Furrow,* January, 1982, p. 32 of Corn Belt edition.
Describes advantages found by two Corn Belt farmers who study why yields of different crop varieties vary so they can apply the answers throughout their operations.

RIDGE TILLAGE

"**No-Till Soybeans Without Herbicides**" by Dick and Sharon Thompson. *The New Farm,* Sept./Oct. 1982, p. 22.
A detailed look at how ridge-planting helps control weeds without chemicals on a 300-acre Iowa farm.

"**More Profit With Less Tillage**" by Ernest E. Behn.
Step-by-step guide to ridge-planting. Available for $9.20 (postpaid) from the author, Rt. 1, Boone, Iowa 50036.

"**Ridging Cuts Costs, Pleases Landlords**" by Mike Brusko. *The New Farm,* Nov./Dec. 1984, p. 34.
Ridge-tillers Donn and Susan Klor describe their first years' experience with a ridging system—the first in their Illinois county.

SOIL FERTILITY

"**A Practical Guide To Novel Soil Amendments**" by Janet C. McAllister, 1983.
Rodale Research Center agronomist reviews more than 80 studies on soil wetting agents, humates, microbial fertilizers and activators, growth regulators, specialty fertilizers and micronutrient products. Available for $6.95 from the Regenerative Agriculture Association, 222 Main St., Emmaus, Pa. 18049.

"**Home-grown Nitrogen**" by Rex Gogerty. *The Furrow,* January/February 1984, p. 16.
Researchers look at soybean and legume varieties for use as homegrown N sources and weed control in crop rotations.

"**A New Clover You Can't Overgraze**" by Fred Zahradnik. *The New Farm,* Nov./Dec. 1981, p. 46.
A report on various subterranean clovers that reseed themselves and grow well alongside pasture grasses.

"**A New Cash Crop Fixes Nitrogen, Too**" by Fred Zahradnik. *The New Farm,* July/Aug. 1982, p. 20.
Describes the success of Idaho farmers who raise chickpeas for profit and homegrown N.

"**Grow Your Own Nitrogen From Seed And Save**" by Fred Zahradnik. *The New Farm,* Nov./Dec. 1982, p. 41.
Describes the advantages of setting aside a few acres for growing your own hairy vetch seed.

"**Fertile Soils Without Chemicals**" Proceedings from the April 1979 conference in Muenster, Saskatchewan.
Talks cover understanding soils, soil salinity, crop rotations, compost, organic research activities in Nebraska, and farm-scale composting. Each presentation followed by extensive question and answer sessions among researchers and farmers. Published by the University of Regina Extension, Regina, Saskatchewan, Canada S4S 0A2.

"**Organic Nutrient Cycling**" by Richard R. Harwood, director of the Rodale Research Center. *The New Farm,* July/Aug. 1980, p. 57.
Discusses how crop rotations, shallower tillage and other practices combine to help release nutrients from soil.

"**Specialty Legumes Maintain Soil Fertility Without Shutting Off Cash Flow**" by Gene Logsdon. *The New Farm,* January 1981, p. 27.
Describes the characteristics of several clovers and vetches and their places in crop rotations.

"**Specialty Legumes**" by John J. Reagan. *The Furrow,* January 1980, p. 18.
Reviews how different types of legumes fit almost every soil condition or livestock need. Mentions strengths of several clovers and vetches, and research done on them.

"**Strategies For Achieving Self-Sufficiency In Nitrogen On A Mixed Farm In Eastern Canada, Based On The Use of the Faba Bean**" by D. Patriquin. In "Genetic Engineering of Symbiotic Nitrogen Fixation and Conservation of Fixed Nitrogen," edited by J. Lyons, et al.; published 1981 by Plenum Co., New York.

"**Soils And Men**" USDA Yearbook of Agriculture, 1938.
Compilation of extensive articles on soil fertility. Yield figures dated, but sound management principles still apply. Out of print but probably in your local library.

"**The Never Never Land Of N**" by George DeVault. *The New Farm,* January 1982, p. 29.
First of the seven-part "Testing...Testing" series on how soil testing labs often produce unreliable fertilizer recommendations. After 69 labs checked identical samples, they recommended from 0 to 210 pounds/acre of nitrogen that could cost up to $31/acre.

"**A Little Phosphorus Goes A Long Way**" by George DeVault. *The New Farm,* February 1982, p. 30.
Second in the "Testing...Testing" series. Labs analyzing identical soil samples recommended from 0 to 150 pounds of P per acre, the cost of which ranges from $0 to $40.

"**Potassium: A Case Of Too Much, Too Often**" by George DeVault. *The New Farm,* March/April 1982, p. 31.
Third of the "Testing...Testing" series. More than half of 70 soil testing labs checking identical soil samples with high K levels recommended an average of 40 pounds/acre more K, which would cost $6/acre.

"**How To Lose $42 An Acre**" by William C. Liebhardt and Martin Culik. *The New Farm,* February 1983, p. 21.
A Rodale Research Center review of fertilizer recommendations shows most labs advise too much NPK—at an average excess cost of $42/acre.

"**Super Alfalfa**" by Fred Zahradnik. *The New Farm,* January 1983, p. 31.
SPREDOR 2, a new creeping variety of alfalfa, can enrich hillside and poor soils while providing high-quality forage at the same time. Covers other special alfalfa varieties.

"**Tougher Than Johnsongrass**" by Fred Zahradnik. *The New Farm,* January 1985, p. 24.
Update on SPREDOR 2 shows creeping roots continue to make this winter-hardy alfalfa an excellent choice for pasture renovation. Offers seeding rates, other details.

WEED CONTROL

"**Controlling Weeds Without Chemicals**" by editors of *The New Farm.*
Describes how plant competition, mechanical weed control and crop rotations can be the key to whipping weeds without herbicides. Available for $2 from the Regenerative Agriculture Association, 222 Main St., Emmaus, Pa. 18049.

"**Ragweed Mowing: Timing Is The Key To Control**" by Bruce Barbour. *The New Farm,* July/Aug. 1981, p. 42.
New Jersey Extension agent reports on study showing that the optimum time to clip ragweed is just prior to flowering.

"**Weed-Free Beans Without Herbicides**" by Dan Looker. *The New Farm,* March/April 1982, p. 14.
Florida soybean grower finds proper planting techniques and cultivation provide successful grass control.

"**Mechanical Weed Control Starts At Planting**" by Larry Mueller. *The New Farm,* February 1982, p. 16.
An Illinois corn and soybean farmer describes how scheduled planting and timely cultivation enabled him to eliminate herbicides.

"**What To Do When They Ban Herbicides**" by Dick and Sharon Thompson. *The New Farm,* January 1984, p. 14.
The Thompsons find soybean varieties that help control weeds without herbicides in their ridge-till system.

"**Seasonality In Organic Weed Management**" by Richard R. Harwood, director of the Rodale Research Center. *The New Farm,* May/June 1980, p. 61.
Explains how understanding a weed's life cycle can improve mechanical control. Examples include pigweed, nutsedge and lambsquarters.

GENERAL

"**Earthcare: Ecological Agriculture In Saskatchewan**" edited by Paul Hanley; published in 1980 by the Earthcare Group.
Introductory manual to ecological farming practices in the prairies. Includes referenced discussion, anecdotal material from organic farmers and good charts. See "Earthcare Wrote The Book On Organic Farming," *The New Farm,* May/June 1982, p. 34. Available for $12.75 (postpaid) from the Earthcare Information Centre, Box 1048, Wynyard, Saskatchewan, Canada S0A 4T0.

"**Problems Facing Canadian Farmers Using Organic Agriculture Methods**" by Dee Kramer, 1984.
Results of 1984 survey of organic farmers throughout Canada. Describes motivations, information sources, production techniques, problems, labor needs, productivity and marketing. Published by Canadian Organic Growers. Available for $4 from the author, 25 Sandbar Willoway, Willowdale, Ontario, Canada M2J 2B1.

"**Alternative Farming Task Force Report**" published 1983 by the University of Nebraska.
Policy paper reviews economics, soil and crop management, soil fertility and nutrient cycling, erosion control, pest management, livestock production, waste management and other concerns associated with organic farming. Available free from the University of Nebraska, Lincoln, Neb. 68503.

"**Environmentally Sound Agriculture**" edited by William Lockeretz, 1983.
Compilation of papers presented at the fourth International Conference on Resource Conserving, Environmentally Sound Agricultural Alternatives, held Aug. 1982 at the Massachusetts Institute of Technology. Subjects include health of crops and livestock, pest management, fertilizers and organic wastes, conservation farming systems, crop quality, organic farming, amaranth, trees and shrubs, biotechnology. Published by Praeger Publishers, New York, N.Y.

"**Updating Organic Farming: New Looks At Old Ways**" by Karl Kessler. *The Furrow*, May/June 1983.
A general look at large-scale organic farms in Illinois, Montana, Minnesota, and Nebraska.

"**Chemicals And Agriculture: Problems and Alternatives**" Proceedings of Nov. 1977 conference in Fort San, Saskatchewan.
Available for $7 from Canadian Plains Research Center, University of Regina, Regina, Saskatchewan, Canada S4S 0A2.

"**The Soul Of Soil — A Guide To Ecological Soil Management**" by Grace Gershuny, 1983.
Explains the structure, character and cycles of soils, and biological factors. Shows how to evaluate soil, make and interpret soil tests. Covers various soil management practices, including drainage, tillage, weeding, manure, compost, green manures, crop rotations. Also covers use of mineral fertilizers. Available for $6.50 (postpaid) from University of Vermont Extension Service, Morrill Hall, Burlington, Vt. 05401.

"**Proceedings Of The Management Alternatives For Biological Farming Workshop**" held in Feb. 1983 at Iowa State University.
Includes discussions on manure management and application; non-chemical control of weeds, plant disease, insects; crop rotation designs; maintenance of soil fertility; and economics of biological production methods. Available for $1 from Robert Dahlgren, Iowa State Cooperative Extension, Wildlife Research Unit, 11 Science 2, ISU, Ames, Iowa 50011.

"**Practical Alternatives To Chemicals In Agriculture**" Proceedings of Oct. 1978 conference in Fort San, Saskatchewan.
Speakers discuss the difficulties encountered when making the transition to "ecological agriculture," manure management, farm-scale composting. Compares studies of organic and conventional farms, green manuring, weed control. Available for $5 from University of Regina Extension, Regina, Saskatchewan, Canada S4S 0A5.

"**Report And Recommendations On Organic Farming**" by USDA, 1980.
Ten-member study team outlines needed research on various non-chemical production methods. Includes case studies of 69 organic farms in 23 states, plus extensive review of organic farming literature. Available for $5.50 from U.S. Government Printing Office, Superintendent of Documents, Washington, D.C. 20402.

"**Organic Farming: Current Technology And Its Role In A Sustainable Agriculture**" Proceedings of 1981 symposium sponsored by the American Society of Agronomy, Crop Science Society of America and the Soil Science Society of America.
Covers comparisons of organic and conventional farming, meeting nutrient needs, reducing pesticide inputs, economics and public policy. ASA Special Publication No. 46. Available for $12 (postpaid) from American Society of Agronomy, 677 South Segoe Rd., Madison, Wis. 53711.

"**Organic Farming**" Series of six reports published in June 1984 in the *Des Moines Register*.
Timely update by Pulitzer Prize-winning farm writer James Risser on current agricultural trends and how organic farming appears at the forefront. Reprints available for $1 from the Regenerative Agriculture Association, 222 Main St., Emmaus, Pa. 18049.

Organic Farming Fact Sheets by William K. Kruesi.
Series of eight reviews focuses on non-chemical weed control in cornfields and pastures, selection and application of natural insecticides, organic fertilizers, green manure crops, seaweed concentrates for horticultural crops, general principles of organic gardening and farming. Available for 25 cents each (postpaid) from University of Vermont Extension Service, 31 The Green, Woodstock, Vt. 05091.

"**How To Survive The Coming Dairy Collapse**" by Vincent Hundt, 1982.
A Wisconsin dairyman argues, in his own 20-page booklet, that farmers should concentrate more on caring for their farms and themselves instead of worrying about the woes of the industry, which he claims can be bypassed. Reviewed in the January 1983 issue of the *The New Farm*. Available for $5 from Hundt, Rt. 1, Coon Valley, Wis. 54623.

A Bibliography For Small And Organic Farmers — 1920-1978 USDA, 1981.
Includes 1,176 publications of long-term research by USDA staff. Selections relate to needs of small and organic farmers. Arranged under 19 categories and year published. Bibliographies and Literature of Agriculture No. 11, available free from U.S. Government Printing Office, Superintendent of Documents, Washington, D.C. 20402.

PEOPLE WHO CAN HELP

Extension and research personnel listed here have a genuine interest in regenerative, resource-efficient agriculture. They'll likely have the information you need, or know where you can find it. Your county Extension agent can be your best source of reliable, research-based information on such items as locally adapted legumes, grasses and grains for green manuring, weed management, and the like. Your agent will also probably know which growers elsewhere in the county have questions similar to yours. And perhaps more than anyone else locally, your district soil conservationist will appreciate what you're trying to do and have practical ideas to support your efforts. In addition, conservation personnel will know the real innovators and land stewards in your region.

CROP ROTATION

Dr. Robert I. Papendick
University of Washington
Pullman, Wash. 99164
(509) 335-1551

—USDA/ARS soil scientist—Grain-legume rotations

Dr. William Hargrove
Georgia Agricultural Experiment Station
Experiment, Ga. 30212
(404) 228-7330

—Assistant professor of soil fertility—Legume cover crops, conservation tillage to minimize fertilizer inputs

Dr. Ronnie R. Duncan
Georgia Agricultural Experiment Station
Experiment, Ga. 30212
(404) 228-7326

—Assistant professor of agronomy—Clover and vetch winter cover crops for cutting nitrogen needs, multiple cropping systems

Anne Duncan Edwards
King George County Extension Service
Rt. 1, Box 1412
King George, Va. 22485
(703) 775-3062

—County agent—Cropping systems, amaranth, allelopathic reactions in crop rotations

GREEN MANURES

Dr. Joseph T. Touchton
Agronomy Dept.
Auburn University
Auburn, Ala. 36830
(205) 826-4100

—Associate professor of soil science—Cover crops, green manures and legume-grain rotations to cut nitrogen fertilizer needs

William Murphy
Crops and Soil Science Dept.
University of Vermont
Burlington, Vt. 05405
(802) 656-3131

—Professor of agronomy—Intensive rotational grazing, pasture renovation

Dr. Wilbur W. Frye
Agronomy Dept.
University of Kentucky
Lexington, Ky. 40546
(606) 257-1628

—Professor of soils—Legumes, overseeding

Dr. Dick L. Auld
Plant and Soil Sciences Dept.
University of Idaho
Moscow, Idaho 83843
(208) 885-7078

—Associate professor of plant breeding—Green manures

Dr. Gary H. Heichel
Agronomy Dept.
University of Minnesota
15059 Gortner Ave.
St. Paul, Minn. 55108
(612) 373-1503

—USDA/ARS professor of plant physiology—Legumes as alternative nitrogen source

Dr. Donald K. Barnes
Agronomy Dept.
University of Minnesota
15059 Gortner Ave.
St. Paul, Minn. 55108
(612) 373-0865

—USDA/ARS professor of alfalfa genetics—Legumes, alfalfa

Dr. Carroll P. Vance
Agronomy Dept.
University of Minnesota
15059 Gortner Ave.
St. Paul, Minn. 55108
(612) 373-1285

—USDA/ARS professor of plant physiology—Legumes

Dr. Craig Scheaffer
Agronomy Dept.
University of Minnesota
15059 Gortner Ave.
St. Paul, Minn. 55108
(612) 373-1677

—Associate professor of agronomy—Forage management in rotations

Dr. Warren W. Sahs
Agronomy Dept.
University of Nebraska Field Lab
Box 72
Wahoo, Neb. 68066
(402) 624-2275

—Field lab supervisor—Legumes, cover crops, rotations

Dr. David F. Bezdicek
Agronomy and Soils Dept.
Washington State University
Pullman, Wash. 99164
(509) 335-3475

—Professor of soil microbiology—Legumes, soil microbiology

HORTICULTURE

Dr. Frank P. Eggert
Plant and Soil Sciences Dept.
University of Maine/Orono
Orono, Maine 04473
(207) 581-2939

—Professor of horticulture—Vegetable crops

INSECT CONTROL

Dr. Stuart Gage
Entomology Dept.
Michigan State University
East Lansing, Mich. 48824
(517) 355-4662

—Environmental systems entomologist—Intercropping fruits and vegetables, non-chemical pest control

Dr. Thomas C. Edens
Entomology Dept.
Michigan State University
East Lansing, Mich. 48824
(517) 355-3346

—Assistant professor of resource economy—Economics of crop rotations, interplanting

Dr. Dean Haynes
Entomology Dept.
Michigan State University
East Lansing, Mich. 48824
(517) 355-4662

—Professor of ecological pest management—Insect population dynamics and impact of new ag designs on them

Dr. George Bird
Entomology Dept.
Michigan State University
East Lansing, Mich. 48824
(517) 355-4662

—Integrated Pest Management specialist—Nematology

Dr. David Pimentel
Entomology Dept.
Cornell University
Ithaca, N.Y. 14853
(607) 256-2212

—Professor of insect ecology—Non-chemical insect control

Dr. John M. Luna
Entomology Dept.
Virginia Polytechnic Institute & State University
Blacksburg, Va. 24061
(703) 961-4823

—Extension specialist—Integrated Pest Management

Dr. Stuart B. Hill
Entomology Dept.
Macdonald Campus of McGill University
P.O. Box 225
Ste. Anne-de-Bellevue
Quebec, Canada H9X 1C0
(514) 457-2000, Ext. 190

—Director, Ecological Agriculture Projects—Biological and integrated pest control

MANURE/COMPOSTING

Vaughn H. Holyoke
Maine Cooperative Extension Service
University of Maine
Winslow Hall
Orono, Maine 04469
(207) 581-2211

—County agent—Manure, composting, crops specialist

Dr. Raymond S. Weil
Agronomy Dept.
University of Maryland
College Park, Md. 20742
(301) 454-4787

—Assistant professor of soil fertility—Soil fertility, organic wastes, composted sewage sludge

Dr. Leon Chesnin
Agronomy Dept.
University of Nebraska
Lincoln, Neb. 68583
(402) 472-1504

—Professor of soil chemistry—Management of livestock, agricultural industrial waste, municipal sludge, large-scale composting; application of organic waste

Les H. Hileman
Agronomy Dept.
University of Arkansas
Fayetteville, Ark. 72701
(501) 575-2354
—Assistant professor of soil fertility—Compost

Stuart D. Klausner
Agronomy Dept.
149 Emerson Hall
Cornell University
Ithaca, N.Y. 14853
(607) 256-2177
—Extension associate of soil conservation—Manure nutrient values, application, maximizing manure's role in crop production

Dr. Parker F. Pratt
University of California at Riverside
Riverside, Calif. 92521
(714) 787-5102
—Professor of soil science and chemistry—Livestock manure, minimum-input farming, rotations

SOIL FERTILITY

Dr. Larry D. King
Soil Science Dept.
North Carolina State University
Raleigh, N.C. 27650
(919) 737-2645
—Associate professor of waste management—Soil chemistry, organic waste management

Dr. Robert Miller
Soil Science Dept.
North Carolina State University
Raleigh, N.C. 27650
(919) 737-2655
—Soil science department chairman—Microbiology of waste disposal, rhizobium ecology, soil fungi

Winston Way
Vermont Cooperative Extension Service
University of Vermont
Burlington, Vt. 05405
(802) 656-2630
—Agronomy specialist—Crops, soils, manure

James Bauder
Montana Cooperative Extension Service
Montana State University
Bozeman, Mont. 59717
(406) 994-5685
—Tillage specialist—Tillage, soil fertility, water quality

Dr. David Patriquin
Biology Dept.
Dalhousie University
Halifax, Nova Scotia, Canada B3H 4J1
(902) 424-2252
—Microbiologist—Nitrogen self-sufficiency, using weeds to farmer advantage

Dr. Guy Mehuys
Soil Science Dept.
Macdonald Campus of McGill University
Ste. Anne-de-Bellevue
Quebec, Canada H9X 1C0
(514) 457-2000
—Professor of agronomy—Green manuring

WEED CONTROL

Dr. Stephen Gliessman
Biological Control of Weeds Laboratory
University of California at Santa Cruz
Santa Cruz, Calif. 95064
(408) 429-4051
—Associate professor of environmental studies—Director of university agroecology program, sustainable agriculture

Dr. Alan Watson
Plant Science Dept.
Macdonald Campus of McGill University
Ste. Anne-de-Bellevue
Quebec, Canada H9X 1C0
(514) 457-2000
—Agriculture faculty researcher—Biological and integrated weed control (knapweeds, thistles, quackgrass, velvetleaf, yellow nutsedge, curly dock)

GENERAL

Patricia Allen
Agroecology Program
University of California at Santa Cruz
Santa Cruz, Calif. 95064
(408) 429-4243
—Outreach coordinator—Farmer liaison between university agroecology program and state Extension

Dr. David J. Sammons
Agronomy Dept.
University of Maryland
Room 1124-A, H.J. Patterson Hall
College Park, Md. 20742
(301) 454-3715
—Associate professor of plant breeding—Intercropping, small grains

In addition, the Research Branch of Agriculture Canada has compiled an updated list of researchers throughout the provinces who are working on various biological insect and weed controls. To obtain a free copy of the list, write.
James S. Kelleher, BioControl Unit, SIRS
Research Program Service, Agriculture Canada
Ottawa, Canada K1A 0C6
Phone: (613) 995-9073

INDEX

Alfalfa, 10-12, 14, 18, 19, 23, 26, 43, 47-52, 57, 59, 71, 77, 78
Allelopathy, 23, 64
Argentine bahiagrass, 62, 63
Armyworms, 63
Austrian winter pea, 10, 11, 22, 26, 67, 68, 73, 76

Banding, 8, 13, 14, 16, 18, 21, 25, 27
Barley, 10, 19, 22, 44, 49, 52, 59, 73, 76
Beggarweed, 83
Beef, 10, 11, 13, 17, 30, 33, 37, 46-48, 52, 57, 58, 62, 68
Beets, 52
Bindweed, 36, 42
Black-eyed peas, 11
Brassicas, 49, 51, 52
Bromegrass, 71
Buckwheat, 7, 12, 49-51
Budworm (tobacco), 70

Careless weed, 83
Chisel plow, 7, 11, 27, 35, 36, 48, 62, 63, 68, 69, 71
Clover, 7, 9, 11-13, 18, 19, 22, 23, 25, 31, 32, 35-39, 41-43, 47-52, 59, 71, 73, 76
Cockelbur, 69
Coffeeweed, 83
Composting, 26, 49, 52
Corn (field), 7-52, 56-60, 64-66, 69-84
Corn (sweet), 55, 56
Cotton, 19, 48, 81, 83
Cover crops, 10, 11, 13, 17, 22-27, 29, 48-52, 68, 71, 73-76. See also Alfalfa; Austrian winter pea; Barley; Brassicas; Buckwheat; Clover; Crop rotation; Double-cropping; Erosion; Fescue; Green manure; Legumes; Nitrogen; Oats; Organic matter; Overseeding; Ryegrass; Seed crops; Small grains; Vetch (hairy); Weed control; Wheat.
Crop rotation, 7-13, 16-26, 30-32, 35, 41, 42, 47, 48, 57-60, 62, 65, 76
Cultivation, 7, 8, 11, 13, 14, 16-18, 20, 21, 25, 27, 28, 33, 35-38, 44, 45, 48, 52, 59, 60, 62, 64-66, 69, 70, 73, 76, 80-84

Dairy, 18, 19, 24, 30, 46-48, 70-72, 76
Disease control (plants), 43, 47, 58, 59, 68, 69, 78. See also Crop rotation; Fungicides; Lentils; Peas (dry); Root rot.
Double-cropping, 8, 22, 24, 25

Earworm, 70
Equipment
 planting, 8, 10, 11, 15, 48, 62-64, 66, 73, 75, 76, 82, 83
 tillage, 8, 11, 27, 48, 62, 64, 66, 69, 71, 76, 81, 82
 weed control, 8, 10, 11, 13-16, 27, 28, 48, 64-66, 69, 73, 80-84. See also Rotary hoe.
Erosion, 7, 10, 15, 22, 23, 26, 58, 62, 67-69, 71, 73-75, 77, 78, 82

Fall panicum, 56
Fertilizer, 7, 9, 10, 12, 13, 15-17, 19-21, 27, 30, 31, 33-35, 37, 38, 42, 44, 47, 49, 55-57, 62, 67, 70, 71, 74, 78-80, 84. See also Greensand; Lime; Magnesium; Micronutrients; Nitrogen; Nutrient deficiencies; pH; Phosphorus; Potassium; Rock phosphate; Soil amendments; Soil testing; Sulfur; Tissue testing.
Fescue, 49-51
Flax, 73
Forages, 9-11, 49-51, 79. See also Alfalfa; Argentine bahiagrass; Austrian winter pea; Brassicas; Bromegrass; Clover; Fescue; Hay; Legumes; Oats; Pasture; Ryegrass; Silage; Sudan grass; Vetch (hairy); Wheat.
Foxtail, 21, 30, 34-37, 48, 56, 64, 65
Fruits, 20, 46, 50, 51, 81
Fungicides, 10, 22, 24, 84

Green manure, 17, 18, 22-24, 29, 30, 36, 48-52, 62, 67, 68, 74-76. See also Alfalfa; Argentine bahiagrass; Austrian winter pea; Barley; Brassicas; Buckwheat; Clover; Cover crops; Crop rotation; Erosion; Fertilizer; Fescue; Legumes; Nitrogen; Oats; Organic matter; Overseeding; Peas (Crowder); Phosphorus; Potassium; Ridging; Ryegrass; Seed crops; Small grains; Vetch (hairy); Weed control.
Greensand, 26

Hay, 7-13, 15, 17, 18, 20-22, 26, 27, 29-31, 35, 37, 38, 41, 46-52, 58-60, 71, 72, 76-78, 81. See also Alfalfa; Austrian winter pea; Black-eyed peas; Bromegrass; Clover; Cover crops; Crop rotation; Erosion; Fescue; Forages; Green manure; Insecticides; Legumes; Orchardgrass; Pasture; Seed crops; Small grains; Sudan grass; Timothy; Weed control; Wheat.
Herbicides, 7-13, 15-22, 24, 25, 27, 28, 30, 31, 33-37, 42, 43, 45, 47, 48, 52, 56-58, 60, 62, 64, 68-70, 75, 77, 80, 82-84.
Hogs, 14, 46-48, 52, 55, 57, 58, 74

Insect control, 10, 16, 18, 19, 30, 33, 43, 47, 49, 50, 51, 59, 63, 69, 70. See also Armyworms; Budworm (tobacco); Earworm; Insecticides; Rootworm; Stink bugs.
Insecticides, 9, 12-14, 16-22, 24, 27, 28, 30, 31, 47, 57, 58, 62, 63, 69, 70, 77, 78, 84

Johnsongrass, 11, 62

Lambsquarters, 36
Legumes, 7-12, 17, 18, 21-24, 27, 29, 30, 32, 49-52, 62, 70, 71, 73-76, 79. See also Alfalfa; Austrian winter pea; Black-eyed peas; Clover; Cover crops; Crop rotation; Double-cropping; Forages; Green manure; Hay; Lentils; Nitrogen; Organic matter; Overseeding; Pasture; Peas; Peas (Crowder); Peas (dry); pH; Seed crops; Soybeans; Vetch (hairy).
Lentils, 52, 68
Lime, 8, 43, 48-51, 71. See also Nutrient deficiencies, pH, Rock phosphate
Livestock, 7, 10, 11, 13, 20, 21, 23, 27-32, 41, 46-48, 52, 78, 84. See also Beef; Dairy; Hogs; Poultry.

Magnesium, 48
Manure, 9-11, 17, 18, 22-24, 26, 27, 30, 33, 34, 37, 38, 41, 43, 47-49, 52, 57-59, 71, 75. See also Composting; Fertilizer; Livestock; Organic matter.
Micronutrients, 15, 50, 51, 69, 76, 80
Millet, 62, 63
Milo (see Sorghum)
Morning glory, 69
Mulch-planting, 8, 81, 82
Mustard, 26, 67

Nematicide, 63
Nitrogen, 7-14, 16-18, 20-25, 27, 29, 30, 32, 34-38, 42, 48-52, 57, 59, 62, 65, 67-69, 73-76, 78-80, 84. *See also* Fertilizer; Legumes.
No-till, 7, 10, 11, 15, 71, 75, 76
Nutrient deficiencies, 32, 34, 37, 38, 45, 47-52, 79

Oats, 18, 19, 23, 27, 31, 35, 36, 38, 39, 42, 43, 47, 49, 51, 52, 59, 71, 73, 77, 78, 84
Orchardgrass, 9-11, 71, 78
Organic matter, 7, 9, 22, 24, 62, 68-71, 73, 75, 80. *See also* Composting; Cover crops; Erosion; Green manure; Manure.
Overseeding, 12, 20, 22-26, 28, 32, 42, 73-76

Pasture, 10, 11, 13, 15, 27, 50, 62, 72, 80
Peas, 52, 73
Peas (Crowder), 62, 63, 68
Peas (dry), 67
pH, 8, 26, 80
Phosphorus, 7, 11, 13-16, 18, 21, 26, 30, 48-52, 55-57, 68, 75, 79. *See also* Austrian winter pea; Fertilizer; pH; Rock phosphate; Soil testing; Tissue testing; Vetch (hairy).
Pigweed, 18, 26, 36, 83
Potassium, 7, 11-16, 18, 21, 26, 30, 48-52, 55-57, 68, 75, 78, 79. *See also* Fertilizer; Greensand; Soil testing; Tissue testing; Vetch (hairy).
Potatoes, 22, 52, 76
Poultry, 46, 57, 58, 68, 74-76

Quackgrass, 18, 71

Ridging, 9, 10, 15, 16, 23, 25, 28, 48, 52, 64-66, 82
Rock phosphate, 7, 14, 17, 18, 26. *See also* Lime; pH; Phosphorus.
Root rot, 78
Rootworm, 9, 13, 21, 24, 27, 33
Rotary hoe, 13, 14, 16, 18, 27, 28, 33, 35, 37, 48, 59, 65, 66, 81, 82
Rye, 44
Ryegrass, 10, 12, 17, 18, 22-24, 47-52, 59, 75, 76

Seed crops, 7, 9, 12, 17, 46, 62, 68, 75, 76. *See also* Argentine bahiagrass; Austrian winter pea; Buckwheat; Clover; Cover crops; Crop rotation; Legumes; Millet; Peas (Crowder); Ryegrass; Vetch (hairy).
Shattercane, 83
Sicklepod, 69
Silage, 11, 18, 30, 31, 33, 35, 39, 59, 60
Small grains, 9, 10, 15, 20, 21, 23, 26, 29, 32, 46-52, 59, 60, 76. *See also* Barley; Buckwheat; Cover crops; Green manure; Oats; Overseeding; Rye; Wheat.
Soil amendments, 18. *See also* Fertilizer; Surfactant; Wetting agent.
Soil testing, 7, 11, 13-16, 18, 26, 31, 44, 55-57, 69, 79
Sorghum, 19, 48, 50, 52, 83
Soybeans, 7-17, 19-22, 24-30, 33-39, 41-44, 46-51, 55, 58-60, 64-66, 74-78, 81-84
Spinach, 55, 56
Stink bugs, 63
Sudan grass, 11
Sulfur, 14, 48, 49, 52
Sugarbeets, 19, 80, 83

Sunflower, 19, 23
Surfactant, 11, 17

Thistles, 42
Timothy, 9, 78
Tissue testing, 14, 15, 34, 36, 38, 45, 56, 75, 79, 80
Tobacco, 46, 50, 51
Tomatoes, 55

Vegetables, 14, 17, 20, 44, 46, 49-51, 74, 75, 81, 83. *See also* Beets; Black-eyed peas; Corn (sweet); Lentils; Peas; Potatoes; Spinach; Sugarbeets; Tomatoes.
Velvetleaf, 18
Vetch (hairy), 7, 12, 13, 17, 22-25, 42, 48-52, 73-76

Weed control, 7-13, 16-18, 20, 21, 23-26, 29, 30, 33-38, 41, 43-45, 47-52, 58-60, 62, 64-66, 68-70, 73, 78, 80-84. *See also* Allelopathy; Banding; Beggarweed; Bindweed; Careless weed; Cockelbur; Coffeeweed; Cover crops; Crop rotation; Cultivation, Fall panicum; Foxtail; Herbicides; Johnsongrass; Lambsquarters; Morning glory; Mustard; Pigweed; Quackgrass; Ridging; Rotary hoe; Shattercane; Sicklepod; Thistles; Velvetleaf; Yellow rocket.
Wetting agent, 17
Wheat, 7-10, 19, 22-24, 26, 30, 35, 37, 42-44, 48, 49, 51, 52, 55, 59, 67, 68, 73, 82, 84

Yellow rocket, 71